Microbiological Risk Assessment Associated with the Food Processing and Distribution Chain

SCIENCES

Agronomy and Food Science, Field Directors –
Jack Legrand and Gilles Trystram

Food Safety, Subject Head – Jeanne-Marie Membré and
Thierry Bénézech

Microbiological Risk Assessment Associated with the Food Processing and Distribution Chain

Coordinated by
Jeanne-Marie Membré

iSTE

WILEY

First published 2022 in Great Britain and the United States by ISTE Ltd and John Wiley & Sons, Inc.

ISTE Ltd
27-37 St George's Road
London SW19 4EU
UK

www.iste.co.uk

John Wiley & Sons, Inc.
111 River Street
Hoboken, NJ 07030
USA

www.wiley.com

Library of Congress Control Number: 2021952704

British Library Cataloguing-in-Publication Data
A CIP record for this book is available from the British Library
ISBN 978-1-78945-084-2

ERC code:
Applied Life Sciences, Biotechnology, and Molecular and Biosystems Engineering
LS9_5 Food sciences (including food technology, food safety, nutrition)
LS9_8 Biohazards (including biological containment, biosafety, biosecurity)

Contents

Preface

Jeanne-Marie MEMBRÉ
SECALIM, INRAE, Oniris, Nantes, France

There are a number of food-associated hazards that can cause harm to human health. Chemicals used in agriculture, environmental pollutants and pathogenic bacteria are all examples of hazards that can ultimately end up on the consumer's plate and in some cases harm their health.

Food safety standards, the development of which began more than 20 years ago, are used to confront these hazards, with their basis in a formal process called "risk analysis". Risk analysis consists of three separate elements: risk assessment, risk management and risk communication.

Three volumes on the subject of Food Safety make it possible to articulate this link between food safety and risk analysis. They have been created by Jeanne-Marie Membré and Thierry Bénézech, with the assistance of Nabila Haddad, and published by ISTE Ltd and Wiley in the SCIENCES series:

– the volume coordinated by Nabila Haddad covers both chemical and microbiological hazards: Haddad, N. (2022), *Hazards in the Food Processing and Distribution Chain*;

– the volume coordinated by Thierry Bénézech and Christine Faille deals with risk management and focuses on microbiological risks: Bénézech, T. and Faille, C. (2022), *Control: Preventing the Biological Risks Associated with Food Contamination During Processing/Distribution and Consumer Usage*;

– this volume, coordinated by Jeanne-Marie Membré, completes the series. It is entitled *Microbiological Risk Assessment Associated with the Food Processing and Distribution Chain.*

Acknowledgements

Jeanne-Marie MEMBRÉ
SECALIM, INRAE, Oniris, Nantes, France

I would first of all like to thank Gilles Trystram and Jack Legrand, who entrusted the joint coordination of the topic of Food Safety to me, together with Thierry Bénézech, and also the publishers and their support team.

I would next like to thank Thierry Bénézech and Nabila Haddad, who assisted me in preparing the three volumes on this topic. A big thank you, too, to the authors of the various chapters, without whom this work would not have been possible.

Finally, all my thanks to Marie-France Pilet, director of SECALIM, the INRAE research unit within which I work, who allowed me the time needed to bring this coordination/editorial project to successful fruition.

Introduction

Jeanne-Marie MEMBRÉ
SECALIM, INRAE, Oniris, Nantes, France

Food safety is the guarantee that food is harmless, in other words that its consumption will have no adverse consequences for health. As we are reminded by the World Health Organization (WHO), food safety encompasses all the measures taken to provide food that is as safe as possible, and hence the policies and measures applied must relate to the entire food chain, from production to consumption.

Ensuring a safe food supply poses major food safety challenges for public authorities (FAO and WHO 2006). Changing global patterns of food production, international trade, the emergence of new technologies, public expectations in terms of health protection and many other factors have created an increasingly demanding environment in which food safety must operate (Figure I.1).

There are a number of food-associated risks that can cause harm to human health. Chemicals used in agriculture, environmental pollutants and pathogenic bacteria are all examples of hazards that can ultimately end up on the consumer's plate and in some cases harm their health.

The first estimates of the global and regional burden of foodborne diseases, published by the WHO in December 2015 (WHO 2015), show that this burden is significant all over the world. Every year around the world, 1 in 10 people fall ill from eating contaminated food and 420,000 die from it, nearly a third of whom are children under the age of 5. The hazards responsible for these foodborne illnesses include bacteria, viruses, parasites, toxins and chemicals. More specifically, diarrheal diseases are responsible for 550 million cases and 230,000 deaths per year. Children are particularly susceptible, with 220 million cases and 96,000 deaths per year. Diarrhea is often caused by eating raw or undercooked meat, eggs, fresh produce and/or dairy produce.

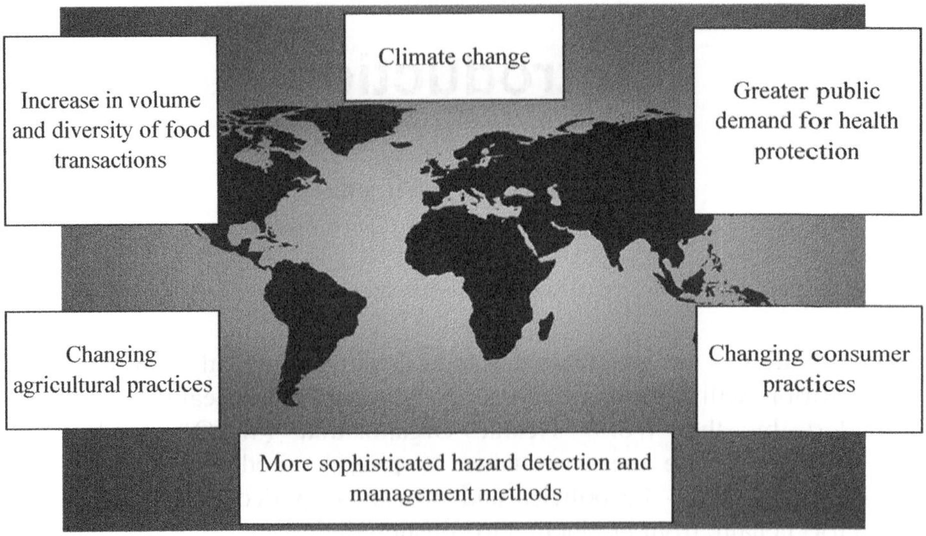

Figure I.1. *Factors influencing the evolution of food safety. From the FAO and WHO (2006)*

In Europe, more than 23 million people fall ill each year after consuming contaminated food, resulting in 5,000 deaths (WHO 2017). Still in Europe, the most common causes of foodborne diseases are noroviruses, followed by *Campylobacter* spp. Non-typhoid *Salmonella* spp. are responsible for the majority of deaths. We will review these three microbiological hazards below.

Norovirus

One of the most common causes of gastroenteritis in humans worldwide is human norovirus, and specifically genogroup 2 noroviruses. Symptoms include diarrhea, vomiting (including projectile vomiting) and stomach pains. The virus can be transmitted to humans through the ingestion of contaminated food or water and directly from person to person, leading to infection in a relatively large proportion of those exposed.

Estimates by the WHO Foodborne Disease Burden Epidemiology Reference Group (FERG) show that, worldwide, norovirus is linked to about 20% of all foodborne illnesses caused by diarrheal disease hazards (125 million per year). It is the most common cause of diarrheal disease for all ages and in all risk groups, with the most serious outcomes being seen in young children and the elderly. Norovirus infection is the sixth leading cause of diarrheal death in children under 5 and the second leading cause of diarrheal death in children over 5, with similar trends in all regions of the world.

According to a study by ANSES (2018), in France, over the period 2006–2015, the main foods associated with outbreaks of collective food toxi-infections related to norovirus were mollusks (72%) and, in particular, oysters (usually consumed raw in France). Composite dishes were implicated in 10% of norovirus outbreaks. The other food categories (meat, fish, vegetables and egg products) were of equal significance and accounted for a total of 18% of outbreaks.

Campylobacter *spp.*

Campylobacter spp. are a major cause of foodborne diarrheal disease. Infection with *Campylobacter* is usually acquired by consuming contaminated foods such as undercooked poultry and raw milk; contaminated broiler meat is considered the most significant source of this hazard to humans.

According to a study by ANSES (2018), in France, over the period 2006–2015, the main food categories identified at the time of *Campylobacter* outbreaks were meats (67%) and composite dishes (18%). Poultry was implicated in 71% of the outbreaks caused by meat. The composite dishes were basically sandwiches, mixed salads and ready-cooked meals. Eggs and egg products, milk and milk products and water accounted for a total of 9% of outbreaks.

Death from campylobacteriosis is rare and is usually limited to very young or elderly patients or those suffering from another serious illness such as AIDS. Complications such as bacteremia, hepatitis, pancreatitis and miscarriage have all been reported with varying frequency. Post-infection complications can include reactive arthritis, which can last for several months, and neurological disorders such as Guillain–Barré syndrome, a form of paralysis that can lead to severe respiratory and neurological dysfunction or death in a few cases.

According to the FERG, in 2010, *Campylobacter* spp. foodborne illnesses were responsible for more than 95 million cases of disease and just over 21,000 deaths worldwide. Since 2005, campylobacteriosis has been the most frequently reported foodborne illness in the European Union, with more than 236,000 confirmed cases in 2014. However, it should be noted that the growing trend in the incidence of campylobacteriosis in recent years is partly due to the improvement of surveillance and diagnostic systems in a number of European Union Member States.

Prevention consists of control measures at all stages of the food chain in order to reduce transmission, from the environment to farms, through enhanced biosecurity, appropriate wastewater and fecal disposal systems, hygienic slaughter and the application of good food-handling practices, in particular avoiding cross-contamination and ensuring proper heat treatment of poultry products.

Non-typhoid Salmonella *spp.*

Non-typhoid *Salmonella* spp. are generally transmitted to humans through contaminated foods, usually of animal origin, such as eggs, pork and poultry, and dairy products; however, other foods, especially fresh produce, may also be involved in their transmission.

Typhoid *Salmonella* spp. generally cause gastroenteritis characterized by the acute onset of fever, abdominal pain, diarrhea, nausea and sometimes vomiting. Infection with *Salmonella* spp. can, however, cause disease, especially in children, the elderly and the immuno-compromised, and can lead to post-infection complications such as reactive arthritis. In the case of invasive *Salmonella* spp., a number of organs and systems can be affected, resulting in bacteremia, meningitis, osteomyelitis or septic arthritis and sometimes even death.

According to FERG estimates, non-typhoidal *Salmonella* spp. are the foodborne hazards responsible for the highest annual burden and the greatest number of deaths both globally and in the European region. Globally, according to FERG estimates, non-typhoid *Salmonella* spp. are responsible for approximately 78 million cases of disease and 59,000 deaths per year.

In Europe, non-typhoid *Salmonella* spp. hold the number one place in terms of deaths due to food risks. For example, in 2014, there were over 85,000 cases of salmonellosis and 65 deaths. The two most frequently reported *Salmonella* serovars in the European Union are *S.* enteritidis and

S. typhi. The European Food Safety Authority has estimated that the overall economic burden of human salmonellosis could reach €3 billion per year.

According to a study by ANSES (2018), in France, over the period 2006–2015, the foods identified at the origin of outbreaks were in 40–45% of cases related to eggs or egg-based preparations. Meats were implicated in around 30% of outbreaks. Milk and dairy products represented only 9% of outbreaks from a suspected or confirmed food, but 19% of outbreaks from a confirmed food. Composite dishes (ready-cooked meals made from multiple ingredients such as couscous, lasagna, pizza, tartiflette) and fishery products were, in total, linked to slightly over 5% of salmonella outbreaks.

Risk analysis

Confronted by these dangers for more than 20 years, the development of food safety standards has been based on a formal process called "risk analysis". Risk analysis includes risk assessment and risk management as well as risk communication (Figure I.2). The Food and Agriculture Organization of the United Nations (FAO) and WHO define risk assessment as a decision-making tool: its aim is not necessarily to extend scientific knowledge, but to provide risk managers with a rational and objective picture of what is known, or assumed to be known, about public health risks and their causes at any given time.

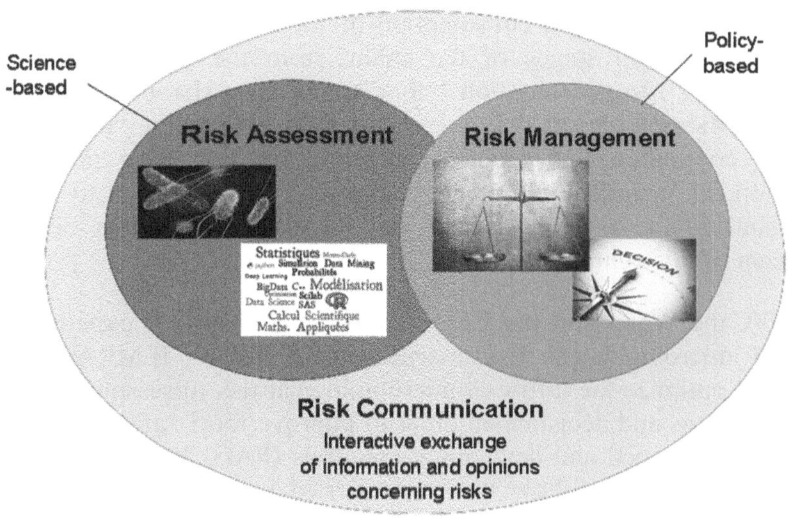

Figure I.2. *The three components of risk analysis: risk assessment, risk management and risk communication. Adapted from the FAO and WHO (2006)*

No matter what the institutional context, the discipline of risk analysis offers a tool that all public authorities can use to improve food safety (FAO and WHO 2020).

This book focuses on risk assessment and more specifically on microbiological risk assessment. Readers interested in chemical risk assessment can refer to a number of books and in particular to the recent book coordinated by Camel *et al.* (2018).

Microbiological risk assessment

The purpose of microbiological risk assessment is to characterize the nature and probability of harm resulting from human exposure to the biological agents present in foodstuffs. Ever since 1999, the WHO and the FAO, through the Codex Alimentarius Commission (CAC), have set out principles and guidelines for conducting microbiological risk assessment of food (Codex Alimentarius Commission 1999). Since then, they have regularly augmented their work; for example, in 2020 they produced a very comprehensive methodological document (FAO and WHO 2020).

Microbiological risk assessment of food is a scientific process comprising four elements (Figure I.3): hazard identification, hazard characterization, exposure assessment and finally risk characterization. In the context of production, distribution and consumption of food products, the risks must take into account all stages of the chain: from the production of raw materials ("farm") to consumption ("fork"), and even through to ingestion ("human"). Microbiological risk assessment thus covers the "farm-to-fork-to-human" continuum. Of course, depending on the food product, its composition, its manufacture and its packaging, the key steps to be taken into account in the risk assessment can be mostly upstream or mostly downstream of this continuum.

In 2000, the WHO and the FAO assembled a group of experts on the subject of microbiological risk assessment: JEMRA. JEMRA aims to develop and optimize the utility of microbiological risk assessment as a tool informing actions and decisions in order to improve food safety and make it available to developed and developing countries (FAO 2020). We strongly recommend the reader follow the work carried out by JEMRA and also consult on a regular basis the WHO and FAO websites dedicated to food safety and microbiological risk assessment.

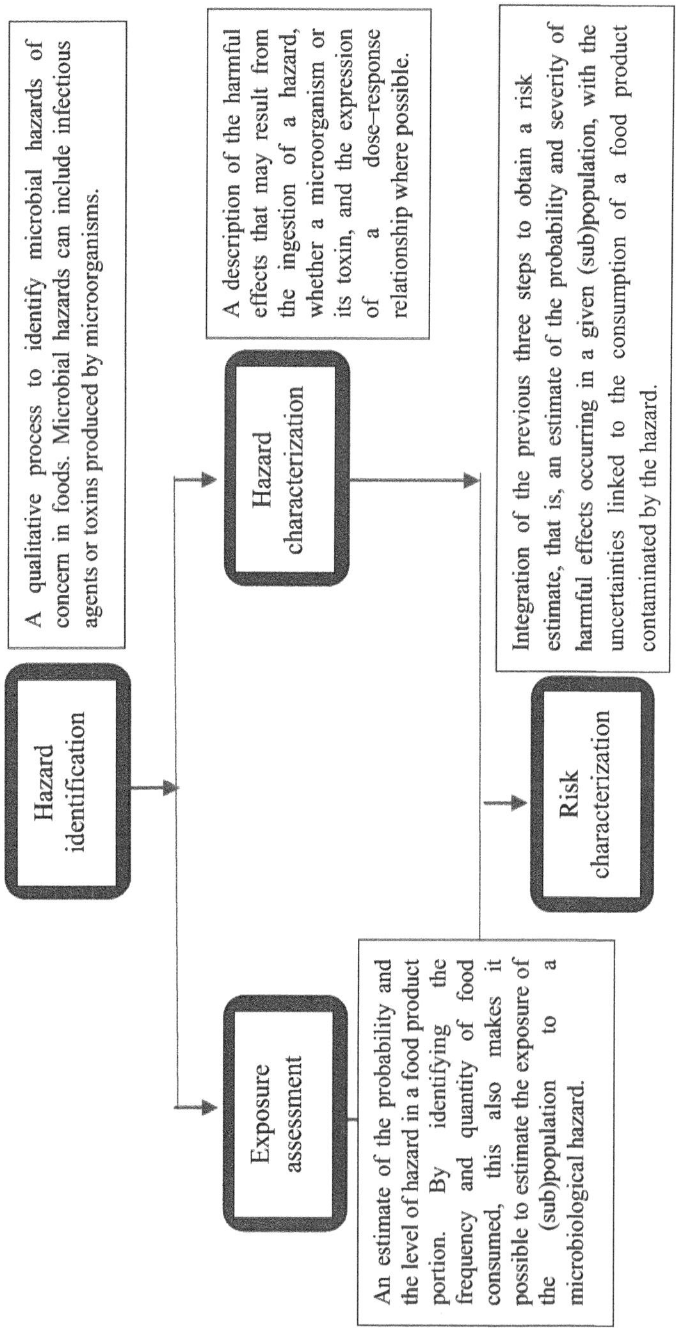

Figure I.3. *Microbiological risk assessment. Adapted from the FAO and WHO (2020)*

The first chapter of this book deals with the first stage of microbiological risk assessment, namely hazard identification. It remains fairly general since the first book in this series is dedicated to hazard analysis (Haddad 2022). The other three stages of microbiological risk assessment, namely hazard characterization, exposure assessment and finally risk characterization, are, however, developed in the various chapters of this book. More specifically, in order to address the biological and mathematical aspects of these steps, hazard characterization and exposure assessment have each been divided into two parts: hazard characterization is subdivided into a pathogenicity mechanism and quantification of the dose–response; exposure assessment is subdivided into methods of detection and enumeration of pathogens and quantification of the level of exposure.

References

ANSES (2018). Avis et rapport de l'Anses relatif à l'attribution des sources des maladies infectieuses d'origine alimentaire. Partie 2 : Analyse des données épidémiologiques. Report [Online]. Available at: https://www.anses.fr/fr/system/files/BIORISK2015SA0162Ra-2.pdf.

Camel, V., Rivière, G., Le Bizec, B. (2018). *Risques chimiques liés aux aliments : principes et applications*. Lavoisier, Paris.

Codex Alimentarius Commission (1999). Principles and guidelines for the conduct of microbiological risk assessment. CAC/GL 30-1999.

FAO (2020). Microbiological risks and JEMRA [Online]. Available at: http://www.fao.org/food/food-safety-quality/scientific-advice/jemra/en/ [Accessed 10 February 2020].

FAO and WHO (2006). Food safety risk analysis: A guide for national food safety authorities [Online]. Available at: http://www.fao.org/3/a-a0822e.pdf. [Accessed November 2021].

FAO and WHO (2020). Public consultation for draft guidance of microbiological risk assessment for food [Online]. Available at: https://www.who.int/news-room/articles-detail/public-consultation-for-draft-guidance-of-microbiological-risk-assessment-for-food [Accessed 15 June 2020].

Haddad, N. (2022). *Hazards in the Food Processing and Distribution Chain*. ISTE Ltd, London and Wiley, New York.

WHO (2015). WHO estimates of the global burden of foodborne diseases. Foodborne Diseases Burden Epidemiology Reference Group [Online]. Available at: https://www.who.int/foodsafety/publications/foodborne_disease/fergreport/en/ [Accessed January 2021].

WHO (2017). The burden of foodborne diseases in the WHO European Region [Online]. Available at: https://www.euro.who.int/__data/assets/pdf_file/0005/402989/50607-WHO-Food-Safety-publicationV4_Web.pdf.

PART 1

Hazards and Food

1

Biological Hazard Identification

Jeanne-Marie MEMBRÉ and Nabila HADDAD
SECALIM, INRAE, Oniris, Nantes, France

1.1. Introduction

Contamination of food with microbial agents is a global public health problem. Microbial hazards in food include bacteria such as *Salmonella*, viruses such as norovirus, parasites such as trematodes and also prions. Diarrheal diseases are the most common diseases resulting from the consumption of contaminated food and are the cause of 550 million illnesses and 230,000 deaths each year (WHO 2020). Hazard identification serves to establish whether the hazard is probable or actual in the food product and to document important known information about the relationships and interactions between the hazard, the food and the host, and also their relationship to human disease. Given that a wide range of microbiological hazards can cause foodborne illness, hazard identification should determine whether a potential hazard is realistic for the food product concerned (FAO and WHO 2020).

Epidemiological data from disease surveillance programs or investigations of food-borne outbreaks are often the first clearly documented indications of a food safety issue associated with a pathogen causing adverse effects. Food contamination surveillance data, along with product and process assessments, can help identify combinations of hazards and foods. Evidence from these sources is usually quantitative (i.e. it includes information on the concentration or number of units of the hazard in the food) and may also provide information that feeds into other stages of microbiological risk assessment, such as exposure assessment and/or establishment of a dose–response relationship. Whole genome sequencing (WGS) is increasingly used for the surveillance of foodborne pathogens, the investigation of epidemics and the search for the sources of contamination throughout the food supply chain (Rantsiou *et al.* 2018).

That being said, the epidemiological data must be cross-referenced with product knowledge, in other words its formulation, its process, its distribution channel and the way it is used by the consumer, before conclusions can be drawn on the relevant hazards. Finally, the behavior of the hazard in the food, in other words its ability to multiply, its resistance to stress and its survival, or its inactivation, is the third key element to be taken into account in hazard identification. (Figure 1.1). We will return later in the chapter to these different types of data and their cross-use.

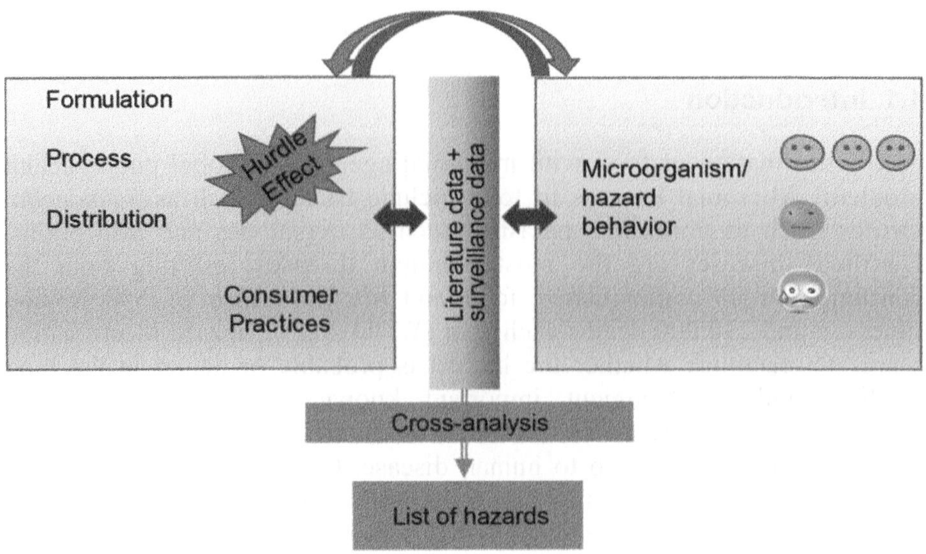

Figure.1.1. *Structuring of information essential to hazard identification. Created with BioRender.com*

1.2. Who conducts hazard identification?

Hazard identification is the first stage in risk assessment. It is therefore often carried out by health agencies such as the European Food Safety Authority (EFSA) in Europe or ANSES in France when these agencies are seeking to clarify an action or a decision relating to food safety.

However, hazard identification is also implemented by companies, or more precisely by companies' quality departments, when they are setting up their Hazard Analysis Critical Control Point (HACCP) systems. The HACCP system is a methodology that identifies, assesses and helps control hazards that have significance for food safety (Membré 2014). The HACCP system is a preventive operation, specific to a manufacturing unit, which begins with the purchase of raw materials, ingredients and packaging materials, follows the entire production process and ends at the finished product, ready for consumption. The HACCP system is based on seven principles, the first being "carry out a hazard analysis". This first principle includes hazard identification. Conducting this identification should be assigned to a highly qualified team with proven skills in food microbiology.

1.3. Sources of useful information for hazard identification

The WHO and the FAO have listed the various types of data that can be used in hazard identification (FAO and WHO 2020), along with their benefits and drawbacks. Their main conclusions are summed up in Table 1.1.

The data cited by the WHO and FAO can be categorized according to their origin (whether they have been obtained from the scientific literature or from surveillance) and nature (epidemiological studies, prevalence and concentrations, early warning, toxi-infection, public health).

Food safety agencies provide professionals in the agrifood sector with factsheets describing biological hazards transmissible through food. These factsheets are updated regularly and are available online; they can also provide useful information for hazard identification. By way of example:

– ANSES' factsheets are available in French at https://www.anses.fr/fr/content/fiches-de-dangers-biologiques-transmissibles-par-les-aliments;

– Canadian factsheets are available at https://www.canada.ca/fr/sante-publique/services/biosecurite-biosurete-laboratoire/fiches-techniques-sante-securite-agents-pathogenes-evaluation-risques.html.

Type of data	Description	Benefits	Drawbacks
Literature data: epidemiological studies	Epidemiological investigations relate to studies that have been commissioned to specifically study the cause and effect relationship between the appearance of foodborne diseases and exposure to certain microbiological risks through the consumption of food.	Epidemiological studies are very specific and provide a vast amount of detailed information about the hazard and the group of consumers studied.	The data are often generated for a relatively small number of consumers and therefore are not representative of larger consumer groups.
Literature data: prevalence and concentration data	Studies identifying the prevalence and count/concentration of target microorganisms at various stages of production/distribution and studies identifying their evolution, such as the effectiveness of a transformation procedure.	These studies are particularly useful for exposure assessment, but can also be used for hazard identification.	The diversity of detection and/or counting methods makes it difficult to compare final estimates. Internationally validated microbiological methods to facilitate this comparison should be preferred.
Surveillance data: early warning systems	A food-safety early warning system allows national authorities to share information on measures taken in response to serious risks detected in relation to food and can thus provide useful information for hazard identification. The European RASFF system is an example of a food-safety early warning system that provides valuable information on product withdrawals and recalls.	An early warning system enables the sharing of data between geographically linked parties in an efficient manner. The data are accessible in a digital format, for example, in the Excel file format in RASFF.	The system is only as good as its least active participant. If one country does not have the resources or the expertise to easily contribute to the data, then the resulting dataset is limited or biased in favor of the other countries in the system. Emerging hazards that are not actively investigated, or that do not need to be reported through a national health system, are less likely to be captured in an early warning system.

Type of data	Description	Benefits	Drawbacks
Surveillance data: collective food toxi-infection data	Appearance of at least two similar cases of a symptomatology, generally gastrointestinal, the cause of which can be attributed to the same food origin. When an infection occurs, an epidemiological investigation is usually undertaken.	Obtaining very specific details about the food source and its preparation in the context of the outbreak is often possible, because of the emphasis on a single food or meal.	There are limitations to using collective food toxi-infection data in preparing dose–response modeling but not really for hazard identification.
Surveillance data: public health data	Countries and several international organizations compile health statistics on infectious/zoonotic diseases. An example of a public health data compilation is Pulsenet.	The data are often very specific, with fairly detailed descriptions of the food (e.g. type, quantity, composition), the pathogen (reliably identified, often subtyped) and the consumer (e.g. age, sex, state of health).	These data are highly dependent on the sophistication of the surveillance system used to collect the information. The data relate only to a limited range of microbiological hazards and do not (or only slightly) reflect sporadic cases.
Surveillance data: food contamination	Governments often have proactive food and water sampling programs in place to detect microbiological hazards, which can be expressed as the percentage of contaminated samples (the prevalence) and/or as concentration in the food. In addition, government agencies (inspection and control departments or designated laboratories) carry out routine surveillance monitoring. These data can be useful for hazard identification and also for exposure or risk assessment.	These activities generate significant amounts of data, in the form of information both on prevalence and on the level of contamination. To permit optimal evaluation of the data on prevalence and level of contamination, the associated meta data should be provided: year, season, geographical location, country, etc.	It may be that these data are not random or fully representative. They are generated in the framework of official control systems that often take resource limitations into account by targeting foods that are known to be problematic.

Table 1.1. *The various types of data that can be used in hazard identification (FAO and WHO 2020)*

In parallel with composing hazard factsheets, ANSES has begun detailed work on source attribution (ANSES 2017a, 2018), which can also be used as a documentary source for hazard identification. Source attribution involves quantifying the relative share of various sources in the burden of infectious diseases transmitted through food (ANSES 2017a). It seeks to allocate the number of cases of infectious diseases transmitted by food to the various sources. Sources include:

– Reservoirs: any living being (human, animal or plant), soil or combination of these upon which a pathogen is primarily dependent for its survival, multiplication and/or reproduction. The reservoir of a pathogen can therefore be animal (e.g. for *Campylobacter* or *Salmonella*), human (e.g. for norovirus or hepatitis A virus), environmental (e.g. for *Bacillus cereus*) or mixed (e.g. for *Listeria monocytogenes* or hepatitis E virus).

– Vehicles: any object or substance which acts as a medium in the transmission of a pathogen from the reservoir to its introduction into a receptive host. In the case of foodborne infections, vehicles are foods contaminated with biohazards (in other words, pathogens or their toxic metabolites). Vehicles can also be environmental (e.g. recreational waters).

A report by ANSES (2018) makes it possible to identify which hazards are linked to which categories of food and/or which sectors. The foods under consideration are milk and dairy products, eggs and egg-based preparations, fishery products, meats, vegetables and composite dishes. For meat, it is possible to narrow it down to the sector level: bovine, porcine, small ruminants, poultry or laying hens. For the fishing industry, a distinction can be made between the fish sector and the mollusk and crustacean sector. This work has been published (Augustin *et al.* 2020) and its chief findings are set out in the Appendices.

Another rich source of information, which is regularly updated, is the documentary collection of the International Commission on Microbiological Specifications for Foods (ICMSF; https://www.icmsf.org/). The sixth book in this series, entitled *Microbial Ecology of Food Commodities*, provides a great deal of information on microbial hazards that can be very useful in a hazard analysis. The book is structured into product categories including meat and meat products, milk and dairy products, poultry products, eggs and egg products, fish and fishery products, cereals and cereal products, vegetables and vegetable products, etc.

With regard to microorganisms and their ability to multiply, their resistance to stress, their survival and their inactivation, the information resulting from validated forecasting models can and must be taken into account. In particular, these models make it possible to quickly assess the potential barrier effects. The scientific literature is full of model and application developments. We recommend the reader look first at summary works (McKellar and Lu 2004; Membré and Valdramidis 2016; Valdramidis *et al.* 2017), starting with Chapter 3 of this book (Exposure Assessment of Microbial Pathogens). In section 1.5, the cook/chill/pH barrier measure against non-proteolytic *Clostridium botulinum* is presented. It is based on the development of a forecasting model applicable to refrigerated, vacuum-packed foods that have received heat treatment sufficient to give them a shelf life (SL) of four weeks or more, with no added preservatives.

Data that are useful for hazard identification, or at least some of these data and assuming the associated metadata (e.g. year, season, geographical location, country) are accessible, can be used in the other stages of microbial risk assessment. As an example, the work carried out by Membré *et al.* (2015) in their study on the risk linked to foie gras is presented in the next section.

1.4. An example of the use of data in hazard identification, exposure assessment and risk characterization

The example developed below is based on the work of Membré *et al.* (2015) on canned foie gras.

Non-acidic canned items require the application of thermal processing to render them healthy, safe and atmospherically stable. The preparation of canned food involves placing the foodstuffs into packaging that is impermeable to gases, liquids and microorganisms, then heating them to a temperature above 100 °C for a variable period of time depending on the food. The temperature charts (time–temperature relationships) that have been used for decades have usually been defined empirically, with the sights set on the main microbiological hazard associated with the manufacture of canned food, proteolytic *Clostridium botulinum*, a sporulating anaerobic mesophilic bacterium (Frazier 1967; Stumbo 1973).

It is generally accepted that a treatment equivalent to at least a temperature of 121.1°C for three minutes (conferring a sterilizing value of three minutes) allows an inactivation of an order of magnitude of 12 log of proteolytic *C. botulinum* spores, and this is the standard for the thermal processing of low-acid foods that are ambiently stable. However, some canned products with an excellent safety history receive a heat treatment below this three minute sterilization value.

For example, the microbiologically safe processing of canned nitritated charcuterie (pâté) relies on mild thermal processing in combination with the presence of nitrite salt in the formulation in order to prevent the growth and formation of toxins from a low level of spores (Anderson *et al.* 2011). Similarly, despite being subject to fairly mild thermal processing, canned foie gras has a proven safety record: no cases of botulism attributed to this product had been reported in the 10 years prior to the study in France (INVS 2013), where 90% of France's output is sold, and moreover, no cases had been reported in the literature (Internet search on www.promedmail.org, www.eurosurveillance.org, www.ncbi.nlm.nih.gov/ and archive.org, which documented more than 700 cases of food botulism from 1918 to 2013).

However, epidemiological data do not provide sufficient evidence to draw firm conclusions on food safety. The authors of the study therefore decided to build a quantitative microbial risk assessment model in order to assess the potential risk associated with proteolytic *C. botulinum* in canned foie gras.

In order to estimate the initial level of contamination of raw foie gras at the stage of their exposure assessment, they used data from manufacturers. Sampling was carried out over nine months with the help of 14 French companies representative of various scales of canning operations. Three hundred and seventy-two raw foie gras samples were taken just before sterilization; they were frozen and sent to the laboratory for analysis. A specific direct count of *C. botulinum* not being possible, a series of steps were carried out. First of all, the count of bacteria forming anaerobic spores was carried out according to the French standard NF V 08-602 (AFNOR 2011) with one modification (a change in the processing temperature). Only the black colonies were counted because they correspond to the Clostridium type. Identification involved partial sequencing of the coding region of 16S rRNA. Finally, for all isolates identified as *C. sporogenes* or *C. botulinum*, specific identification of the *C. botulinum* genes was carried out according to a variety of protocols (Raphael and Andreadis 2007; Fach *et al.* 2009; Woudstra *et al.* 2012).

To estimate the level of risk affecting the French population and to use that to validate the estimated level of risk during the study, epidemiological data covering the period 2001–2012 were collected and analyzed. In France, information on botulism epidemics is compiled each year by the National Institute for Public Health Surveillance (INVS 2013). The data for the 12 years were extracted and then analyzed to estimate the number of cases per million inhabitants in France.

The quantitative risk assessment model estimated, with current consumption, an average probability of illness of 8.0×10^{-10} per inhabitant and per year. This value is very low, which led to the conclusion that current practices in terms of the thermal processing of canned foie gras were sufficient to control the risk of the presence of *C. botulinum* in foie gras in France.

It is worth noting here that this study was supported by the *Comité Interprofessionnel des Palmipèdes à Foie Gras*, a body that brings together all professional stakeholders in the sector. It was carried out by scientists from the academic and professional spheres, not by health authorities. This supports what was mentioned in section 1.2: hazard identification and risk assessment are tools that can be used by a wide-ranging scientific community.

1.5. The hazard identification process

The example developed with foie gras is quite revealing of the approach to hazard identification, to which we now return in more detail.

First of all, precise information must be gathered on the product, its ingredients, its manufacture (decontamination steps, possible recontamination in the factory), its storage method (frozen, refrigerated, ambient) and its method of use (for immediate consumption after opening or for use over a period of time).

Then, attention must be focused on the barrier effect put in place to make it fit for consumption (healthy and wholesome). If, for example, the product is canned and sterilized after packaging, there is no need to focus on vegetative cells that are incapable of withstanding the high-temperature thermal processing (sterilization means a scale of 121.1°C for at least three minutes, or its equivalent). Likewise, if the product is derived from lactic fermentation then for foods with an acidic pH distributed under refrigeration for consumption within a few weeks, such as natural yogurt in France, there is no need to focus on pathogenic bacteria such as *Salmonella*, the proliferation of which (if present initially) is inhibited by the double-barrier effect of acidity and storage at refrigerated temperature. These two products being, moreover, logically consumed at one sitting, the risk of recontamination on the consumer's premises is very low. This first rapid analysis, based on the barrier effects put in place, can then be cross-referenced with epidemiological data (Table 1.1) before reaching a firm conclusion on the list of hazards to be retained for the assessment of the potential risk associated with these two types of products.

We mentioned above the use of the end product by the consumer. Unless the epidemiological data (literature or surveillance) indicate the contrary, it is logical to carry out hazard identification taking into consideration consumer practices as stated on the packaging or as reported from consumer surveys. It is futile to consider the "worst-case scenario" as the central point of analysis. For example, in the case of yogurt, it is reasonable to assume its immediate consumption in the minutes after opening. On the other hand, the consumer is not one entity: there are many consumers and many behaviors. This variability in behavior must therefore be taken into consideration when carrying out hazard identification, let alone risk assessment. To illustrate this last point, we will take the example of mayonnaise.

The mayonnaise sold in supermarkets and hypermarkets (homemade mayonnaise is excluded from this) is packaged in France either in a tube or in a jar and stored at the point of sale at room temperature in the grocery section. Before its purchase, this product is stable, and this stability is due to its processing (mainly pasteurized ingredients), its formulation (acetic acid, NaCl, etc.) and its very nature, the emulsion reducing the amount of free water (ICMSF 2005). Before purchasing the product, the barrier effect is therefore threefold: the processes, formulations and nature of the product. On the other hand, once the product is purchased, it is not consumed all in one go but, on the contrary, in several goes. It says on the packaging "to be kept in a cool place and consumed within the month", but mayonnaise is typically a product that the consumer takes a few liberties with: some consumers keep it in a cool place, others at room temperature (put away with the oil, vinegar and condiments because it is associated with this type of product), some consume it in the week or the month but others consume the whole jar or tube regardless of its initial date of opening (which they have in any case forgotten). None of these behaviors, whether occurring frequently or less frequently, can be ruled out right from the start during the hazard identification stage. The risk of recontamination on the consumer's premises exists and is even fairly high when the mayonnaise is packaged in a jar: in the middle of the table during the meal, the diners serve and re-serve themselves as they wish. This is thus the reason why in a so-called "industrial" mayonnaise, pathogenic bacteria of the *Salmonella* type, pathogenic *Escherichia coli* and *Listeria monocytogenes* must be taken into consideration. The barrier effect must be sufficient to guarantee their inactivation in the event of recontamination; this effect is based on the emulsion (quantity of oil of at least 65% in the United States) and the formulation: a pH less than 4 in the United States, acetic acid (which, beyond its effect on the pH, has an inhibitory effect, like most weak acids) and addition of preservatives, such as sorbic acid (ICMSF 2005). There is no doubt that consumer demand for natural products without preservatives and/or reduced-fat products will change the formulation of this product, and therefore hazard identification of hazards should be reviewed in line with these changes.

It has been said above that hazard identification requires precise information be gathered on the product, its ingredients, its manufacture, its method of storage and its mode of use. This information must be cross-referenced with epidemiological data, whether sourced from the literature or from national surveillance plans, and with knowledge of the behavior of the hazard (Figure 1.1). Sliced ham, sold in supermarkets, is a very good illustration of the cross-referencing of these three types of information. In fact, the production diagram (set out in the Appendices) shows that there is a cooking process that is sufficient to eliminate the vegetative forms of pathogenic bacteria. On the other hand, this same diagram shows that there is a handling stage (namely slicing) after cooking, which is a possible source of recontamination. At the same time, epidemiological data (literature and surveillance) point to a significant percentage of ham slices contaminated with *Listeria monocytogenes* during marketing. Finally, knowledge of the behavior of the hazard tells us that *L. monocytogenes* is quite capable of multiplying in cold conditions. Consequently, in the hazard identification process, *L. monocytogenes* must be taken into consideration. This example, developed in the Appendices, is taken from an article published in 2016 based on data analyzed in 2015 (Zwietering *et al.* 2016). The data have not been updated, so there may be a discrepancy between the figures presented and current figures, but the reasoning remains the same.

One final example of hazard identification is linked to refrigerated, vacuum-packed foods that receive heat treatment sufficient to give them a SL of four weeks or more, with no added preservatives. However, this thermal processing is still relatively mild in order to allow these products to keep all their qualities of texture and flavor. This balance is maintained by using knowledge of the main hazard (psychrotrophic *Clostridium botulinum*) and a specific barrier effect called the degree of protection (DoP). An example concerning vacuum-packed refrigerated products and the DoP principle is presented here. More details are provided in the Appendices. This example is sourced from the work of Membré *et al.* (2009).

Clostridium botulinum is one of the most serious hazards considered in food safety. There are well-established control procedures that have been identified to destroy or inhibit the growth of non-proteolytic spores of *C. botulinum.* One of these, used in the manufacture of refrigerated foods, is the application of "non-proteolytic *C. botulinum* cooking", based on the fact that spores from non-proteolytic strains are considerably more heat-sensitive than spores from proteolytic strains. It is thus accepted that a temperature of 90°C for 10 minutes results in an inactivation of 6 $\log_{10}$ of non-proteolytic *C. botulinum* (Gould 1999) and this rule is commonly used for the thermal processing of refrigerated foods. However, in order to maintain the texture and flavor of refrigerated products such as vegetable puree, it is possible to try to drop below this bar ("safe harbor") of 90°C for 10 mins. In this case, some spores of *C. botulinum* could survive. However, under stress conditions provided by intrinsic factors (e.g. pH) or extrinsic factors (e.g. storage at refrigerated temperatures), the surviving heat-treated spores require some time (referred to here as the "latency period") before they recover from their injury, germinate and start to develop.

The objective of this study was therefore to explore the possibility of identifying heat treatments milder than 10 minutes at 90°C, integrating this latency time with the classic thermal inactivation effect. An exposure assessment model, based on the concept of the DoP, was developed.

The principle of the approach used to calculate the DoP, combining the thermal inactivation effect and the thermal damage effect, is based on similar studies previously presented in the literature (Hauschild 1982; Lund 1993; Schillinger *et al.* 1996). Details are provided in equations [1.1]–[1.4]:

$$\mathrm{DoP} = \Delta\mathrm{R} + \mathrm{ThI} \quad [1.1]$$

with ΔR being the decimal reduction due to thermal inactivation and ThI the DoP due to thermal stress. Both are expressed as the decimal logarithm of the reciprocal of a probability:

$$\Delta\mathrm{R} = \log(1/\mathrm{Pr}) \quad [1.2]$$

$$\mathrm{ThI} = \log(1/\mathrm{Pi}) \quad [1.3]$$

Pr indicates the probability that a spore will survive thermal treatment and Pi the probability that the latency period will be shorter than the SL:

$$\mathrm{Pi} = \mathrm{Probability}\ (\mathrm{Lag} \leq \mathrm{SL}) \quad [1.4]$$

Once the model has been built, the results can be set out in two ways. The first is to present the DoP, including thermal inactivation and stress, for different processes, formulations (pH and aw) and refrigerated storage conditions at different shelf lives.

The second is to choose a targeted level of protection and present the model results as a set of combinations of processes, formulations, storage conditions and SLs that achieve that target protection. This second presentation of the results can be compared to an isoprobability method, because the DoPs are derived from probability calculations. The second approach was the one chosen in this study. An overall DoP of 6 was chosen because it is a reference value currently applied in the management of food safety, in the form of inactivation (Gould 1999). Work by Membré *et al.* (2009), which led to this approach, along with bibliographical references and a number of illustrations are presented in the Appendices.

It was shown that the hazardous *C. botulinum* was controlled in a product with a pH of 6.0 kept in a cold place for 25 days, despite the thermal processing being reduced to 85°C for 10 minutes. These results were obtained by taking into account the potential variability of refrigeration temperatures at the premises of both the distributor and the consumer. Probabilistic approaches were implemented. Such approaches are discussed again in the chapter on exposure assessment and in the chapter on risk characterization.

1.6. Scientific work on hazard identification

A Web of Science search (January 2021) using the query "TITLE: (hazard AND identification) AND TOPIC: (microbial OR microbiological OR microorganism*)" first of all revealed the very small number of scientific articles in the field of hazard identification. Indeed, since 1950, there have been only 61 articles which directly address (in the sense that the keywords are in the title) the topic of microbiological hazard identification.

Once the off-topic articles had been discarded, and focusing on the most frequently cited articles (10 times or more), only 16 articles were left (Table 1.2). Their detailed analysis sheds valuable light on both the most frequent object of study and the new trends/methods used in hazard identification.

Hazard identification was most often done as part of HACCP systems or risk (or exposure) assessment, which is consistent with what has already been said in section 1.2. Finally, the articles reporting on molecular approaches were published from 2015. The two cited (Franz *et al.* 2015; Pielaat *et al.* 2015) were identified in the literature search because the term "hazard identification" was in the title of the article, but there are other articles that deal with the use of molecular methods and in particular with WGS in hazard identification; see, for example, the review by Rantsiou *et al.* (2018).

As noted in the Introduction, hazard identification is the first step in microbiological risk assessment. This chapter does not develop examples; they are merely provided to illustrate the process. The reader is invited to consult the first book in this series (Haddad 2022) to see numerous examples of hazard identification.

Title of publication	Reference	Publication Year	Scope
Hazard identification in swine slaughter with respect to foodborne bacteria	Borch *et al.* (1996)	1996	Application of hazard identification
Live bacterial vaccines – A review and identification of potential hazards	Detmer and Glenting (2006)	2006	Application of hazard identification
An approach to reduction of salmonella infection in broiler chicken flocks through intensive sampling and identification of cross-contamination hazards in commercial hatcheries	Davies and Wray (1994)	1994	Food safety management
A comparison between broiler chicken carcasses with and without visible fecal contamination during the slaughtering process on hazard identification of *Salmonella* spp.	Jimenez *et al.* (2002)	2002	Food safety management
An identification procedure for foodborne microbial hazards	van Gerwen *et al.* (1997)	1997	General approach
Identification, assessment and management of food-related microbiological hazards: Historical, fundamental and psycho-social essentials	Mossel *et al.* (1998)	1998	General approach
Hazard identification and exposure assessment for microbial food safety risk assessment	Lammerding and Fazil (2000)	2000	General approach
Identification of hazards and critical control points (CCP) for cassava fufu processing in south-west Nigeria	Obadina *et al.* (2008)	2008	HACCP
Akara ogbomoso: Microbiological examination and identification of hazards and critical control points	Lateef *et al.* (2010)	2010	HACCP

Title of publication	Reference	Publication Year	Scope
Identification of the risk factors associated with cheese production to implement the hazard analysis and critical control points (HACCP) system on cheese farms	Carrascosa *et al.* (2016)	2016	HACCP
First step in using molecular data for microbial food safety risk assessment; hazard identification of *Escherichia coli* O157:H7 by coupling genomic data with *in vitro* adherence to human epithelial cells	Pielaat *et al.* (2015)	2015	New approach: the use of molecular techniques
Molecular hazard identification of non-O157 Shiga toxin-producing *Escherichia coli* (STEC)	Franz *et al.* (2015)	2015	New approach: the use of molecular technique
Microbiological hazard identification and exposure assessment of street food vending in Johannesburg, South Africa	Mosupye and von Holy (2000)	2000	Risk or exposure assessment
Microbiological hazard identification and exposure assessment of food prepared and served in rural households of Lungwena, Malawi	Taulo *et al.* (2008)	2008	Risk or exposure assessment
Hazard identification and exposure assessment for bacterial risk assessment of informally marketed milk in Abidjan, Cote d'Ivoire	Kouame-Sina *et al.* (2012)	2012	Risk or exposure assessment
Hazard identification and characterization: criteria for categorizing Shiga toxin-producing *Escherichia coli* on a risk basis	Asakura *et al.* (2019)	2019	Risk-based approach

Table 1.2. *List of 16 scientific papers dealing with the identification of the most frequently cited hazards (based on a search of the Web of Science database in January 2021)*

1.7. Conclusion and perspective

Hazard identification is, and will remain, an essential step in food safety. The work of hazard identification requires cross-referencing different types of knowledge and information:

– knowledge of the product and its processes, provided in particular by field expertise;

– knowledge of consumer practices, provided in particular by surveys, in addition to labeling, which indicates the main trends;

– knowledge of microbial behavior provided by previous studies from the scientific literature, hazard factsheets prepared by health safety agencies, forecasting models, etc.;

– epidemiological data including surveillance data on prevalence and concentrations, withdrawals/recalls indicated by early warning systems, reports of toxi-infections at the country level, public health data, etc.

This entails a substantial amount of research work , as well as, above all, documentary work. This work should be assigned to a highly qualified team with proven skills in food microbiology.

Several avenues of research in hazard identification can be envisaged for the future. One, which is not new as such but which will definitely take on importance, is the early identification of emerging hazards constrained by change. Kleter and Marvin (2009) have proposed three types of constraints classified as those related to changes in the environment, in the production methods (processes and formulations) of the food and in the food preferences of consumers.

The environment is changing, especially under the influence of climate change, which will affect the emergence and persistence of bacteria, viruses, parasites, harmful algae, fungi, etc. The WHO is expecting projected climate change to negatively impact on food safety, especially in low- and middle-income countries (WHO 2019). Even in Europe, climate change will have consequences for food safety. For example, a recent review focusing on microbial risks in the dairy industry (Feliciano *et al.* 2020) pointed out that raw milk is sensitive to climate change, as its quality can be affected by fluctuations in ambient temperature and the amount of rainfall. Climate change can induce microbiota changes and stress in lactating cows, with a risk of increasing their vulnerability to microbial infections and consequently higher microbial contamination of milk.

Under its Food 2030 agenda, the European Union considers food safety as part of a food system including nutrition, climate, circularity (resource efficiency) and innovation (empowering communities) (European Commission 2016). In the EU, there are societal expectations for sustainable food production, especially in terms of food safety and quality and also environmental and animal welfare standards (European Commission 2016). This much more general integration of food safety in a study of a complete food system (from new cultivation methods to consumer expectations) opens up perspectives in risk assessment which are taken up again in the Conclusion of this book. This is also important to take into account when identifying hazards since it can lead to the emergence or re-emergence of new hazards: new ingredients, new combinations of ingredients, new cooking calculations, new processes, etc.

Regarding consumer food preferences, a recent report in the *Sénat* (Cartron and Fichet 2020) pointed to a boom in France in organic and sustainable agricultural production methods. The turnover from organic food has multiplied by 10 in 20 years, with an acceleration since the early 2010s. Likewise, the consumer is turning to locally sourced products. ANSES (2017b) has flagged up other changes in the eating practices of French consumers; for example, they are tending to eat more raw products, consume seeds or nuts on a regular basis and cut back on dairy products and replace them with soy products. These practices are followed in the best of faith, but may present certain health risks. A recent study conducted at INRAE (Eygue *et al.* 2020) has shown that these new practices generate a wide variety of hazards, such as bacteria, parasites, mycotoxins, allergens and chemical compounds, which are potentially linked to a wide range of health impacts, such as food poisoning, anaphylaxis, cancer, immunosuppression and endocrine disorders.

Hazard identification is therefore a "never-ending story". It must be reviewed and augmented regularly on the basis of new knowledge and new tools made available to the scientific community and must take into account the perpetual changes in the environment and in the production and consumption of food products.

1.8. References

AFNOR (2011). Norme NF V08-602 – Microbiologie des aliments – Dénombrement des spores dans les produits alimentaires avant traitement d'appertisation par comptage des colonies. AFNOR Store Publishing, Paris.

Anderson, N.M., Larkin, J.W., Cole, M.B., Skinner, G.E., Whiting, R.C., Gorris, L.G.M., Rodriguez, A., Buchanan, R., Stewart, C.M., Hanlin, J.H., Keener, L., Hall, P.A. (2011). Food Safety Objective approach for controlling *Clostridium botulinum* growth and toxin production in commercially sterile foods. *Journal of Food Protection*, 74(11), 1956–1989.

ANSES (2017a). Attribution des sources des maladies infectieuses d'origine alimentaire. Partie 1 : Revue des méthodes et inventaire des données [Online]. Available at: https://www.anses.fr/fr/system/files/BIORISK2015SA0162Ra.pdf.

ANSES (2017b). INCA 3 : Evolution des habitudes et modes de consommation, de nouveaux enjeux en matière de sécurité sanitaire et de nutrition Anses – Agence nationale de sécurité sanitaire de l'alimentation, de l'environnement et du travail [Online]. Available at: https://www.anses.fr/fr/content/inca-3-evolution-des-habitudes-et-modes-de-consommation-de-nouveaux-enjeux-en-mati%C3%A8re-de.

ANSES (2018). Avis et rapport de l'Anses relatif à l'attribution des sources des maladies infectieuses d'origine alimentaire. Partie 2 : Analyse des données épidémiologiques [Online]. Available at: https://www.anses.fr/fr/system/files/BIORISK2015SA0162Ra-2.pdf.

Asakura, H., Boisen, N., Chinen, I., Cook, R., Dallman, T., Devleesschauwer, B., Feng, P., Franz, E., Fratamico, P., Gill, A., Griffin, P., Keddy, K., Mainda, G., Majowicz, S., Pires, S.M., Ogunrinola, Y., Scheutz, F., Srimanote, P., Alvarez, R.V., Esteban, E., Lejeune, J., Mylrea, G., Moreira, D., Shaw, W.K., Cahill, S., Ouattara, B., Desmarchelier, P., Nakagawa, R., Zhou, K., Grp, F.W.S.E. (2019). Hazard identification and characterization: Criteria for categorizing Shiga toxin-producing *escherichia coli* on a risk basis. *Journal of Food Protection*, 82(1), 7–21.

Augustin, J.-C., Kooh, P., Bayeux, T., Guillier, L., Meyer, T., Jourdan-Da Silva, N., Villena, I., Sanaa, M., Cerf, O., Blanchemanche, S., Bonnaud, L., Gautier, M., Gauchard, F., Nabec, L., Soler, L.-G., ANSES Working Group (2020). Contribution of foods and poor food-handling practices to the burden of foodborne infectious diseases in France. *Foods*, 9(11), 1644–1661.

Borch, E., Nesbakken, T., Christensen, H. (1996). Hazard identification in swine slaughter with respect to foodborne bacteria. *International Journal of Food Microbiology*, 30(1–2), 9–25.

Carrascosa, C., Millan, R., Saavedra, P., Jaber, J.R., Raposo, A., Sanjuan, E. (2016). Identification of the risk factors associated with cheese production to implement the hazard analysis and critical control points (HACCP) system on cheese farms. *Journal of Dairy Science*, 99(4), 2606–2616.

Cartron, F. and Fichet, J.-L. (2020). Vers une alimentation durable : un enjeu sanitaire, social, territorial et environnemental majeur pour la France. Report, Sénat, Paris.

Davies, R.H. and Wray, C. (1994). An approach to reduction of *Salmonella* infection in broiler chicken flocks through intensive sampling and identification of cross-contamination hazards in commercial hatcheries. *International Journal of Food Microbiology*, 24(1–2), 147–160.

Detmer, A. and Glenting, J. (2006). Live bacterial vaccines – A review and identification of potential hazards. *Microbial Cell Factories*, 5, 23–34.

European Commission (2016). Food 2030 – European R&I for Food and Nutrition Security [Online]. Available at: https://doi.org/doi:10.2777/069319.

Eygue, M., Richard-Forget, F., Cappelier, J.-M., Pinson-Gadais, L., Membré, J.-M. (2020). Development of a risk-ranking framework to evaluate simultaneously biological and chemical hazards related to food safety: Application to emerging dietary practices in France. *Food Control*, 115, 107279.

Fach, P., Micheau, P., Mazuet, C., Perelle, S., Popoff, M. (2009). Development of real-time PCR tests for detecting botulinum neurotoxins A, B, E, F producing *Clostridium botulinum*, *Clostridium baratii* and *Clostridium butyricum*. *Journal of Applied Microbiology*, 107, 465–473.

FAO and WHO (2020). Public consultation for draft guidance of microbiological risk assessment for food [Online]. Available at: https://www.who.int/news-room/articles-detail/public-consultation-for-draft-guidance-of-microbiological-risk-assessment-for-food [Accessed 15 June 2020].

Feliciano, R.J., Boué, G., Membré, J.-M. (2020). Overview of the potential impacts of climate change on the microbial safety of the dairy industry. *Foods*, 9(12), 1794.

Franz, E., Van Hoek, A.H.A.M., Wuite, M., Van Der Wal, F.J., De Boer, A.G., Bouw, E.I., Aarts, H.J.M. (2015). Molecular hazard identification of non-O157 Shiga toxin-producing *Escherichia coli* (STEC). *PloS One*, 10(3).

Frazier, W.C. (1967). *Food Microbiology*. McGraw-Hill Inc., New York.

Gould, G.W. (1999). Sous vide foods: Conclusions of an ECFF Botulinum working party. *Food Control*, 10(1), 47–51.

Haddad, N. (2022). *Hazards in the Food Processing and Distribution Chain*. ISTE Ltd, London and Wiley, New York.

Hauschild, A.H.W., Hilsheimer, R., Jarvis, G., Raymond, D.P. (1982). Contribution of nitrite to the control of *Clostridium botulinum* in liver sausage. *Journal of Food Protection*, 45(6), 500–506.

ICMSF (2005). *Micro-organisms in Foods 6: Microbiological Ecology of Food Commodities*, 2nd edition. Kluwer Academic and Plenum Publishers, New York.

INVS (2013). Caractéristiques épidémiologiques du botulisme humain [Online]. Available at: http://www.invs.sante.fr/Dossiers-thematiques/Maladies-infectieuses/Risques-infectieux-d-origine-alimentaire/Botulisme/Donnees-epidemiologiques [Accessed 06 August 2014].

Jimenez, S.M., Salsi, M.S., Tiburzi, M.C., Pirovani, M.E. (2002). A comparison between broiler chicken carcasses with and without visible faecal contamination during the slaughtering process on hazard identification of *Salmonella* spp. *Journal of Applied Microbiology*, 93(4), 593–598.

Kleter, G.A. and Marvin, H.J.P. (2009). Indicators of emerging hazards and risks to food safety. *Food and Chemical Toxicology*, 47(5), 1022–1039.

Kouame-Sina, S.M., Makita, K., Costard, S., Grace, D., Dadie, A., Dje, M., Bonfoh, B. (2012). Hazard identification and exposure assessment for bacterial risk assessment of informally marketed milk in Abidjan, Cote d'Ivoire. *Food and Nutrition Bulletin*, 33(4), 223–234.

Lammerding, A.M. and Fazil, A. (2000). Hazard identification and exposure assessment for microbial food safety risk assessment. *International Journal of Food Microbiology*, 58(3), 147–157.

Lateef, A., Davies, T.E., Adelekan, A., Adelere, I.A., Adedeji, A.A., Fadahunsi, A.H. (2010). Akara ogbomoso: Microbiological examination and identification of hazards and critical control points. *Food Science and Technology International*, 16(5), 389–400.

Lund, B.M. (1993). Quantification of factors affecting the probability of development of pathogenic bacteria, in particular *Clostridium botulinum*, in foods. *Journal of Industrial Microbiology*, 12, 144–155.

McKellar, R.C. and Lu, X. (2004). *Modeling Microbial Responses in Food*. CRC Press, London.

Membré, J.-M. (2014). HAZARD APPRAISAL (HACCP). Establishment of performance criteria. In *Encyclopedia of Food Microbiology*, 2nd edition, Batt, C. and Tortorello, M.-L. (eds). Elsevier, Amsterdam.

Membré, J.-M. and Valdramidis, V. (2016). *Modeling in Food Microbiology: From Predictive Microbiology to Exposure Assessment*. ISTE Press, London, and Elsevier, Oxford.

Membré, J.-M., Wemmenhove, E., McClure, P.J. (2009). Exposure assessment model to combine thermal inactivation (log reduction) and thermal injury (heat-treated spore lag time) effects on non-proteolytic *Clostridium botulinum*. In *Safety, Reliability and Risk Analysis: Theory, Methods and Applications*, Martorell, S., Soares, C.G., Barnett, J. (eds). CRC Press and Taylor and Francis Group, London.

Membré, J.-M., Diao, M., Thorin, C., Cordier, G., Zuber, F., André, S. (2015). Risk assessment of proteolytic *Clostridium botulinum* in canned foie gras. *International Journal of Food Microbiology*, 210(0), 62–72.

Mossel, D.A.A., Weenk, G.H., Morris, G.P., Struijk, C.B. (1998). Identification, assessment and management of food-related microbiological hazards: Historical, fundamental and psychosocial essentials. *International Journal of Food Microbiology*, 40(3), 211–243.

Mosupye, F.M. and Von Holy, A. (2000). Microbiological hazard identification and exposure assessment of street food vending in Johannesburg, South Africa. *International Journal of Food Microbiology*, 61(2–3), 137–145.

Obadina, A.O., Oyewole, O.B., Sanni, L.O., Tomlins, K.I., Westby, A. (2008). Identification of hazards and critical control points (CCP) for cassava fufu processing in south-west Nigeria. *Food Control*, 19(1), 22–26.

Pielaat, A., Boer, M.P., Wijnands, L.M., Van Hoek, A.H.A.M., El, B., Barker, G.C., Teunis, P.F.M., Aarts, H.J.M., Franz, E. (2015). First step in using molecular data for microbial food safety risk assessment: Hazard identification of *Escherichia coli* O157:H7 by coupling genomic data with *in vitro* adherence to human epithelial cells. *International Journal of Food Microbiology*, 213, 130–138.

Rantsiou, K., Kathariou, S., Winkler, A., Skandamis, P., Saint-Cyr, M.J., Rouzeau-Szynalski, K., Amézquita, A. (2018). Next generation microbiological risk assessment: Opportunities of whole genome sequencing (WGS) for foodborne pathogen surveillance, source tracking and risk assessment. *International Journal of Food Microbiology*, 287, 3–9.

Raphael, B.H. and Andreadis, J.D. (2007). Real-time PCR detection of the nontoxic nonhemagglutinin gene as a rapid screening method for bacterial isolates harboring the botulinum neurotoxin (A-G) gene complex. *Journal of Microbiology Methods*, 71, 343–346.

Schillinger, U., Geisen, R., Holzapfel, W.H. (1996). Potential of antagonistic microorganisms and bacteriocins for the biological preservation of foods. *Trends in Food Science and Technology*, 7(5), 158–164.

Stumbo, C.R. (1973). *Thermobacteriology in Food Processing*, 2nd edition. Academic Press, New York.

Taulo, S., Wetlesen, A., Abrahamsen, R., Kululanga, G., Mkakosya, R., Grimason, A. (2008). Microbiological hazard identification and exposure assessment of food prepared and served in rural households of Lungwena, Malawi. *International Journal of Food Microbiology*, 125(2), 111–116.

Valdramidis, V., Cummin, E., Van Impe, J. (2017). *Quantitative Tools for Sustainable Food and Energy in the Food Chain*. Eurosis Publisher, Ghent.

Van Gerwen, S.J.C., De Wit, J.C., Notermans, S., Zwietering, M.H. (1997). An identification procedure for foodborne microbial hazards. *International Journal of Food Microbiology*, 38(1), 1–15.

WHO (2019). Food safety, climate change and the role of the WHO [Online]. Available at: https://www.who.int/foodsafety/publications/all/climate_change/en/ [Accessed January 2021].

WHO (2020). Assessing microbiological risks in food [Online]. Available at: https://www.who.int/activities/assessing-microbiological-risks-in-food [Accessed 10 February 2020].

Woudstra, C., Skarin, H., Anniballi, F., Fenicia, L., Bano, L., Drigo, I., Koene, M., Bäyon-Auboyer, M.H., Buffereau, J.P., De Medici, D., Fach, P. (2012). Neurotoxin gene profiling of *Clostridium botulinum* types C and D native to different countries within Europe. *Applied and Environmental Microbiology*, 78, 3120–3127.

Zwietering, M.H., Jacxsens, L., Membré, J.-M., Nauta, M.J., Peterz, M. (2016). Relevance of microbial finished product testing in food safety management. *Food Control*, 60, 31–43.

2

Detection and Enumeration of Foodborne Pathogens and Bacterial Toxins in Food

Nabila HADDAD
SECALIM, INRAE, Oniris, Nantes, France

2.1. Why should we look for pathogens in food?

2.1.1. *Foodborne infectious diseases: a major economic and public health issue*

The World Health Organization (WHO) estimates that 1.5 million people died from diarrheal diseases worldwide in 2005. A separate study found that foodborne illnesses cause 70% of diarrheal diseases, mainly due to pathogenic bacteria and viruses (Käfferstein 2003). Norovirus, hepatitis E virus, *Salmonella* spp., *Campylobacter* spp., *Listeria monocytogenes*, *Staphylococcus aureus* and pathogenic *Escherichia coli* are the main pathogens that cause the highest numbers of outbreaks linked to food sources (Dewey-Mattia *et al.* 2018; Holland and Mahmoudzadeh 2020).

Scallan *et al.* (2011) estimated that 31 major pathogens acquired in the United States caused 9.4 million episodes of foodborne illness, leading to 55,961 hospitalizations and 1,351 deaths annually (Scallan *et al.* 2011). Over 95% of these illnesses, hospitalizations and deaths due to known pathogen causes are caused by only 15 pathogens. In the European region, more than 22 million diarrheal foodborne diseases and almost 3,000 deaths are

Microbiological Risk Assessment Associated with the Food Processing and Distribution Chain,
coordinated by Jeanne-Marie MEMBRÉ. © ISTE Ltd 2022.

estimated per year (Kruse 2015). Diarrheal diseases account for 96% of foodborne illnesses and 62% of deaths (Kruse 2015). In addition to the numbers of cases, hospitalizations and deaths, another indicator of the burden of foodborne diseases is disability-adjusted life years (DALYs): these represent the number of years of potential life lost due to premature mortality and the years of productive life lost due to disability. Kirk *et al.* (2015) estimated that 22 foodborne infectious diseases resulted in 25.2 million DALYs due to contaminated food (Kirk *et al.* 2015). The burden greatly varies across the world; the highest incidence and the highest death rates are found in the African and southeast Asian regions.

Every illness has an economic cost. Foodborne diseases bring about huge financial costs at different levels, that is, communities, individual nations and the whole world. For example, the economic burden of the 15 leading foodborne illnesses in the United States was estimated to be more than $15.5 billion in 2013 (Hoffmann *et al.* 2015). This economic analysis included the medical treatment costs of non-hospitalized cases and the costs of chronic illnesses and disabilities, of hospitalizations and deaths and of productivity losses and the willingness to pay to reduce the risk of additional deaths (Hoffmann *et al.* 2015). The analysis underestimated the total cost of foodborne infectious diseases because more than 200 microorganisms can be responsible for foodborne illness (Mead *et al.* 1999). Underestimation is accentuated because most of these illnesses are not reported to the health authorities; moreover, the cause is often unidentified when illnesses are declared.

Other countries have also estimated the annual costs of foodborne diseases, and global estimates of their burden are currently being developed. Snowdon *et al.* (2002) speculated that the economic consequences of foodborne diseases are substantial in both industrialized and developing nations (Snowdon *et al.* 2002), although the consequences may be greater in developing countries because of the high incidence of infant mortality, malnutrition and chronic diarrhea.

The economic and public health burdens of foodborne diseases and the main characteristics of the main foodborne pathogens are discussed in Chapter 3.

2.1.2. *Regulations and establishment of microbiological criteria*

The general principles with regard to feed and food law applicable in the European Union (EU) are stated in the General Food Law, Regulation (EC) 178/2002, the aim of which is to protect public health and consumers' interest in relation to food. The regulation applies to all stages of the production, processing and distribution of food and feed, except domestic activities. In addition, this regulation also defines the structure and tasks of the European Food Safety Authority (EFSA).

According to regulation 178/2002, "it is illegal to place food on the market if it is unsafe. Food is unsafe if it is considered:

– injurious to health;

– unfit for human consumption" (European Commission 2002).

Microbiological criteria provide guidance on the acceptability of foodstuffs and their manufacturing processes. Preventive actions such as the application of good hygiene and good manufacturing practices (GHPs, GMPs) and the Hazard Analysis Critical Control Point (HACCP) principles contribute to achieving food safety. Microbiological testing alone cannot guarantee the safety of a foodstuff, but these criteria provide objectives and reference points to assist food businesses and competent authorities in their activities to manage and monitor the safety of foodstuffs, respectively.

Microbiological criteria are either mandatory (standards) or advisory (guidelines and specifications). A mandatory criterion may not exceed the specified limit, including rejection, destruction or reprocessing. An example of a mandatory criterion (standard) is the "zero tolerance" set for *Salmonella* spp. in minced meat and meat preparations (European Commission 2007).

Various organizations, such as the Food and Agriculture Organization (FAO), World Health Organization (WHO), the Codex Alimentarius International Food Standards (CAIFS) and the International Commission on Microbiological Specifications of Foods (ICMSF), have been involved in developing general principles and applications of microbiological criteria. Scientific advice on matters related to microbiological risks in food is provided by the EFSA. Several categories of criteria are employed for application purposes in the food industry:

– standards refer to mandatory microbiological criteria that are a part of a law, ordinance or administrative regulation. They should specify limits for pathogenic microorganisms of public health significance;

– guidelines refer to microbiological criteria that are often used by the food industry or regulatory agencies to monitor a manufacturing process. They provide mechanisms aimed at checking whether microbiological conditions at critical control points (CCPs) are within the normal range. They are used to assess processing efficiency at CCPs and compliance with GMPs. A microbiological guideline also outlines hygienic significance;

– specifications are used as microbiological criteria in purchase requirements of food ingredients between a seller and buyer. A microbiological specification can also be applied at a specified point during or after processing to monitor hygiene. It is used to warn the manufacturer.

European Commission regulation 2073/2005 on microbiological criteria for foods, applicable from January 1, 2006, states food safety criteria for relevant foodborne bacteria, their toxins and metabolites, such as *Salmonella*, *L. monocytogenes*, *Cronobacter sakazakii*, staphylococcal enterotoxins and histamine, in specific foods (European Commission 2007). These criteria define the acceptability of a product or a batch of food as applicable to marketed products. In addition, this regulation specifies certain process hygiene criteria for the correct functioning of the production process.

European regulation 2073/2005, amended by regulation 1441/2007 on the microbiological criteria applicable to foodstuffs, defines two types of microbiological criteria (European Commission 2005, 2007). Safety criteria define the acceptability of a product or a batch of food. They are applicable to marketed products until the end of their shelf life. The process hygiene criteria specify the acceptability of the production process. Such criteria are not applicable to marketed products. They set an indicative value for contamination above which corrective measures intended to maintain the hygiene of the process are required, in accordance with food legislation, but they do not conclude on the compliance or non-compliance of a product. Criteria other than regulatory microbiological criteria are established by professionals on the basis of hazard analysis, sector by sector, and on a voluntary basis.

2.1.3. *Food safety and quality control*

To guarantee the safety of food products and to prevent unsafe products reaching the consumer, agri-food companies have to implement a plan to control food safety risks. To check that these risks are properly controlled throughout the manufacturing process, they must especially control facilities, raw materials and finished products by themselves. As soon as they become aware of unfavorable results for these finished products, they must inform the public authorities and proceed with the withdrawal/recall of these foodstuffs.

Moreover, the competent authorities of the EU countries arrange official controls of the food industry to verify business compliance with the requirements of the agri-food chain legislation. As for the state services, they are responsible for monitoring the application of food safety regulations by professionals to ensure compliance with legislation specifying maximum levels of particular pathogens in certain categories of food products (European Commission 2007). They implement monitoring plans to assess the contamination levels of food products intended for human consumption by substances that may pose a risk to the consumer's health.

Consequently, the presence of specific bacteria and their concentrations must be determined so as to assess and control safety hazards and the potential for spoilage and ensure correct product characteristics. The most commonly used forms of bacteriological analysis in food microbiology are detection and enumeration. For that purpose, methods and rapid tests have been developed and improved to monitor foods for the presence of foodborne bacterial pathogens.

Inexpensive and user-friendly, culture-based methods are historically considered to be the gold standard (Bhunia 2014). However, these methods require at least two to three days to yield results and must generally be followed by other complementary assays (such as biochemical or serotyping tests) to confirm that the isolate is indeed the pathogen of interest. Due to the perishable nature and therefore the limited shelf life of many foods, the long duration of microbiological analysis makes these tests inadequate in many cases. In order to detect or even quantify foodborne pathogens in food, and to overcome the limitations of culture-based tests, several alternative and generally faster culture-independent methods have been optimized or developed. Jasson *et al.* (2010) proposed to define "alternative methods" as any method or system with "a shorter time to detection" and/or "a better flow through/handling of multiple samples", with such methods thus entailing "convenience and automation in work in the lab" (Jasson *et al.* 2010).

2.1.4. *Definitions of methods*

Quantitative or qualitative microbiological methods can be used for food analysis. Quantitative methods or enumeration are used to determine the number of bacteria directly or indirectly present in a given sample. Qualitative methods aim at demonstrating whether the target agent is present in the sample. The methods used to detect and isolate bacterial pathogens have evolved over the years from cold-enrichment techniques to conventional and molecular methods (WHO and World Organization for Animal Health 2014).

Conventional culture-based methods are usually considered as reference methods for food microbial analysis. Nonetheless, other types of methods (e.g. molecular-based methods) can be recognized as reference methods. Current ISO methods for detecting enterohemorrhagic *E. coli* or *L. monocytogenes* are among them.

This chapter summarizes and categorizes some of the methods used for microbiological analysis of food and highlights their advantages and limitations. This is not an exhaustive analysis of all the existing methods used for the microbiological analysis of foodstuffs, but only a focus on the most common, the most documented and/or the most promising methods.

2.2. Conventional culture-based methods: how to detect and quantify pathogenic bacteria in food

2.2.1. *Isolation and colony-count methods*

Traditional culture relies on the ability of bacteria to grow and multiply on laboratory media (broth or agar media) and form visible colonies or turbidity or display color changes. Methods based on culture assays are designed to enable the detection of a single target cell in a sample. The relatively low number of bacteria present in a food sample limits the sensitivity of all the various types of tests available to evaluate food safety, including those based on culture.

Since pathogenic germs are considered fragile and often found in low numbers in food products, it is important to increase the chances to detect them. Factors such as heat, freezing, freeze-drying, drying, salts, preservatives and other chemicals or natural antimicrobial compounds can affect pathogens in foods (Wu 2008). In addition, bacteria under a viable but non-culturable (VBNC) state can be induced in response to food environmental stresses (Oliver 2010) and make their recovery critical during culturing procedures.

For this purpose, culture-based methods employ selective agents and enrichment-plating techniques. Selective agents inhibit other competing microflora, whereas enrichment increases the number of pathogens up to detectable levels along with the recovery of injured or stressed cells. The pre-enrichment step in a non- or half-selective medium revivifies and increases the population of injured target microorganisms. This step leads to the dilution of inhibiting compounds and rehydration of bacterial cells derived from dried or processed food matrices (Dwivedi and Jaykus 2011). The second enrichment step involves a selective medium (containing different salts and antibiotics) in order to eliminate the background microflora and increase the numbers of the target pathogen. This results in a million-fold multiplication of the target, allowing for its isolation or detection (Jasson *et al.* 2010).

When a culture-based method is used, the detection of bacteria requires inoculation on solid-selective media from the enrichment medium or from the concentrated suspension. Presumed positive colonies are isolated on a selective differential agar medium. Isolation media vary depending on the germs that are sought for. Detecting the suspected pathogen (appearance of typical colonies on the medium) or obtaining a negative result (no typical colony) may take four days. Additional morphological, biochemical, physiological and/or serological testing confirms the presumed colonies. Confirming a positive result can take around one week (Jasson *et al.* 2010; Dwivedi and Jaykus 2011). Furthermore, pure colonies of the targeted bacteria are suitable for epidemiological surveillance and outbreak management purposes (WHO and World Organization for Animal Health 2014).

Chromogenic and fluorogenic isolation media make it easier to recognize colonies of presumed target pathogens. These microbiological growth media contain substrates linked to a chromogen (color reaction), a fluorogen (fluorescent reaction) or a combination of the two. The target bacterial population is characterized by enzyme systems that metabolize the substrate, leading to the release of the chromogen/fluorogen. This leads to a color change in the medium and/or fluorescence under long-wave UV light. Some chromogenic media (e.g. agar *Listeria* according to Ottaviani and Agosti

(ALOA) and tryptone-bile-glucuronic (TBX) agar for *E. coli*) have therefore been taken up in the standardized ISO methods (ISO 11290-1/A1:2004 and ISO 16649:2001, respectively).

One of the quantitative microbiological methods is the colony count method. The bacterial concentration present in a food sample can be determined via two methods: (i) the spread-plate method that consists of inoculating dilutions onto the surface of a solid growth medium, or (ii) the pour-plate method, which requires mixing the test portion with the liquefied agar medium in Petri dishes. Enumeration is performed after incubation for fixed time periods under temperature and atmospheric conditions that vary according to the targeted pathogen (from 7 to 55 °C in an aerobic, micro-aerobic or anaerobic atmosphere). Classical culture methods have a quantification limit of ca. 4 colony-forming units (CFUs)/mL for liquid foods or ca. 40 CFUs/g for solid foods.

Bacterial concentrations can be estimated using the most probable number (MPN) method in food samples suspected to have low contamination (Bell *et al.* 2005). Dilutions of food samples are prepared in the same way as in the plate count method. After incubation, the number of positive tubes for each dilution is counted. An MPN in the original sample is determined based on this number, using standard MPN tables and taking the dilution factor into account.

A variety of rapid culture-based methods have been developed without necessarily shortening the detection time. These methods are designed to reduce the workload and facilitate the workflow by reducing handling and/or the need for a complete laboratory infrastructure. Some of these modified culture methods are based upon the colony-count method (e.g. 3M® Petrifilm®; Bird *et al.* 2020), whereas others apply the principle of the MPN method (e.g. TEMPO®; Owen *et al.* 2010).

2.2.2. *Illustration of the culture-based method:* Listeria monocytogenes *case study*

Three internationally recommended culture-based methods are commonly used for isolation of *L. monocytogenes* on/in a wide variety of food matrices (Figure 2.1). These methods include the ISO international standard 11290 method, the United States Department of Agriculture (USDA) method and the

One Broth *Listeria* method approved by the Association Française de Normalisation (AFNOR) (Zhang *et al.* 2004; Gasanov *et al.* 2005; Gómez *et al.* 2014). These methods must be applied in accordance with their scope and can be food product-specific. The ISO standard 11290 is recommended for isolating *L. monocytogenes* in a large variety of food and feed products as well as environmental samples (Liu *et al.* 2007). The USDA method is recommended for detecting *Listeria* spp. in meat and poultry products and environmental swabs (Ponniah *et al.* 2012). The One Broth *Listeria* method is used for isolating *Listeria* spp. from dairy food products, meat, seafood and vegetables (Gasanov *et al.* 2005).

The ISO 11290 and USDA methods require a series of primary and secondary enrichments of the samples in selective broth media, whereas the One Broth *Listeria* method involves a one-step enrichment that takes two days to produce results as opposed to four to five days for the other two methods (Figure 2.1) (Leong *et al.* 2017). These enrichment media contain different selective agents aimed at mostly inhibiting the growth of Gram-negative bacteria that are often present as competitors in food samples. The mechanisms of inhibition vary among these antibiotics. They include the inhibition of protein synthesis (cycloheximide, fosfomycin and nalidixic acid), disruption of the bacterial outer cell membrane (colistin) and beta-lactamase (cefotetan), among others.

Following the enrichment steps, selective media are used to plate the enriched broth. Oxford and PALCAM agar are required by ISO 11290, while the chromogenic media like ALOA and RAPID'L.mono are recommended by the USDA (Leong *et al.* 2017). The One Broth Listeria method uses *Listeria* brilliance green agar. These media are typically dependent on the β-glucosidase activity of *Listeria*, which cleaves the chromogenic substrate and produces blue or green colonies. In addition, lecithin present in agar is hydrolyzed by a phospholipase only synthesized by *L. monocytogenes*, leading to the formation of opaque halos around their colonies (Ponniah *et al.* 2012). Presumed *L. monocytogenes* colonies on selective agar are confirmed by complementary tests based on bacterial biochemical properties (Gram staining, catalase testing, motility testing, the ability to cause hemolysis on blood agar plates, etc.).

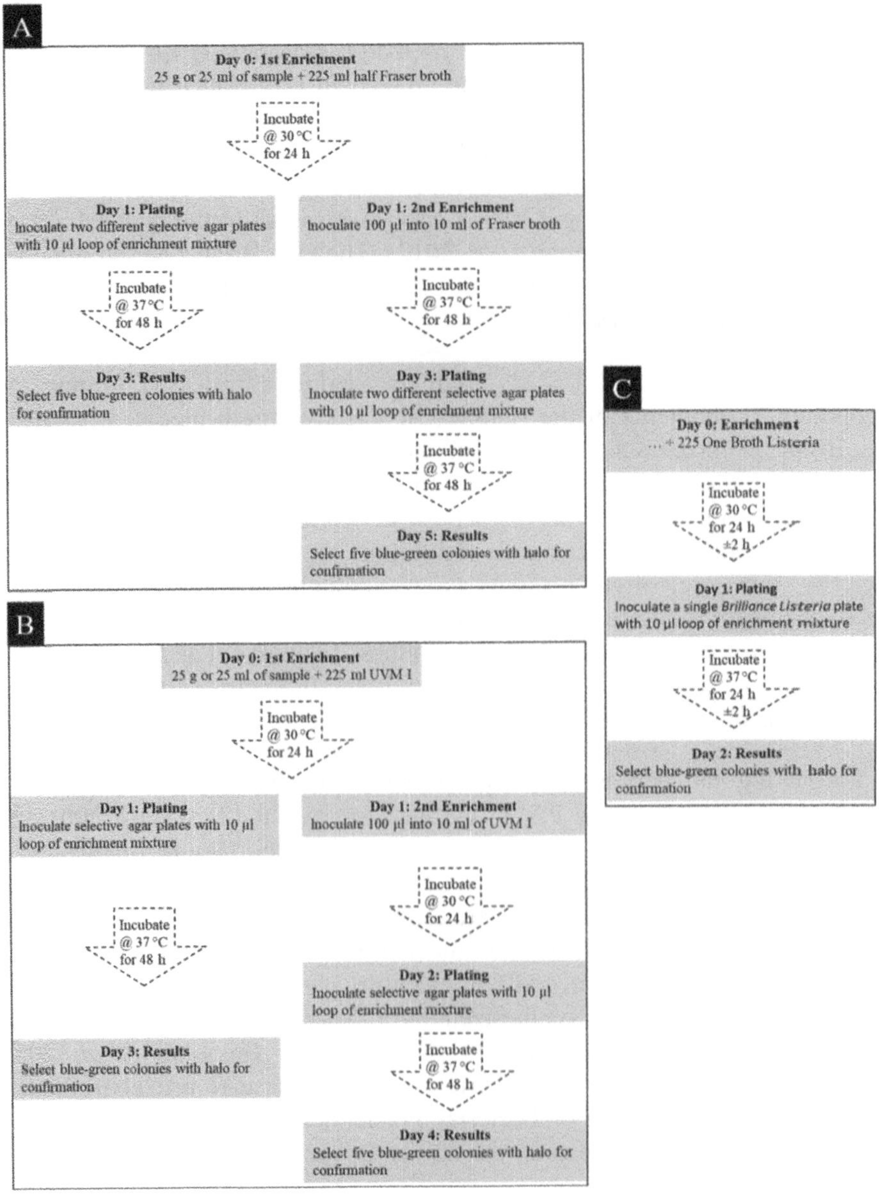

Figure 2.1. *Schematic representation of the three internationally recommended culture-based methods for* Listeria monocytogenes *isolation: (A) ISO standard 11290 (B) the United States Department of Agriculture (USDA) method and (C) the One Broth Listeria method. Adapted from Leong* et al. *(2017) and Matle* et al. *(2020). UVM I: University of Vermont media*

The main advantages of culture-based methods for food microbiological analysis include sensitivity, easy handling, cost-effectiveness and the fact that they remain the "gold standards" compared with other validated methods. Additionally, these methods are the top choice for laboratories because they give qualitative or quantitative information on the number and type of viable microorganisms present in the food samples.

Among the drawbacks of these methods, we can list low bacterial strain resolution, and phenotypic changes because of environmental selection (Välimaa *et al.* 2015; Leong *et al.* 2017). In addition, false-negative results can ensue from contaminating bacteria or from atypical reactions due to atypical strains (Välimaa *et al.* 2015; Leong *et al.* 2017). Moreover, these methods are generally tedious: they require several sequential steps before definitive identification (Bhunia 2014). The entire culture process typically takes two to three days for preliminary isolation and up to one week for the final confirmation of the species isolate. The accuracy of the results depends on (i) the non-uniform distribution and often low abundance of pathogens in a food sample, (ii) the heterogeneity of food matrices and (iii) the presence of an indigenous microflora (Mandal *et al.* 2011). Culture-based methods can also have limited detection capabilities if the targeted pathogen is in an injured or VBNC state in the food.

Molecular-based methods were developed to overcome some of these limitations when detecting and quantifying pathogens present in food samples. These methods can share some steps with culture-based methods (Figure 2.2).

2.3. Nucleic acid-based methods

2.3.1. *Conventional polymerase chain reaction-based methods*

Nucleic acid-based methods operate by detecting specific DNA or RNA sequences of the target pathogenic organism. The emergence of the polymerase chain reaction (PCR) technique changed the way microbiological testing was done by detecting specific microbial DNA as a target. The intense development of this technique allowed the emergence of more advanced methods like quantitative real-time PCR (qPCR) and digital PCR (dPCR). These techniques offer the possibility of reliably diagnosing and quantifying pathogens in food.

2.3.1.1. *Sample preparation*

Generic information on PCR-based methods as given by Vidic and Auger (2019) is summarized below.

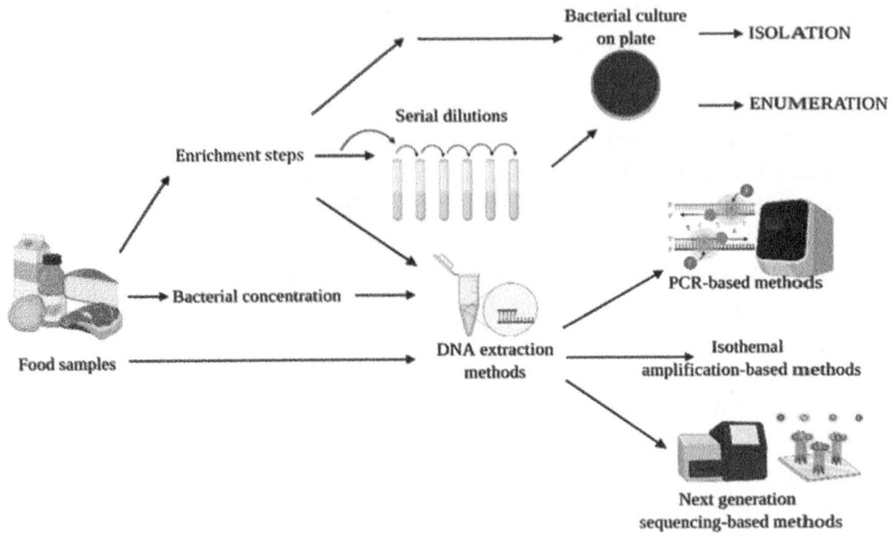

Figure 2.2. *Conventional culture-based and nucleic acid-based methods for the detection of pathogenic bacteria in food. Culture-based methods consist of detecting and/or enumerating the targeted bacteria on selective media. Molecular methods are based on the extraction of bacterial DNA followed by polymerase chain reaction (PCR), isothermal amplification or partial or total DNA sequencing. Adapted from Vidic and Auger (2019). Created with BioRender.com*

Molecular methods based on the use of nucleic acids require extracting and purifying bacterial DNA. DNA extraction can be preceded by sample enrichment as in culture-based methods, in order to revive and multiply bacteria before their detection by PCR (Figure 2.2). Another advantage of this enrichment step is the dilution and elimination of PCR inhibitors (salts, preservatives, etc.) present in the sample. DNA extraction from culture media is easier than direct extraction from food samples, which are much more heterogeneous in terms of composition (Margot *et al.* 2015).

The pre-enrichment step can be replaced by a pre-concentration step of the targeted pathogens. Although this procedure can be laborious and expensive, pre-concentration may be necessary to obtain sufficient DNA and detect bacteria present in low numbers in a sample. Filtration or centrifugation concentrates the bacterial cells of a sample based on their molar masses and dimensions. This procedure is particularly relevant for liquid food such as milk or water.

It is important to note that the enrichment and pre-concentration steps are the main contributors to the cost of bacterial analysis of food.

The ISO 21571:2005 standard method was developed for the extraction of bacterial DNA from food matrices. It provides general requirements as well as specific methods for the extraction, purification and quantification of DNA. DNA quality and purity are critical factors for PCR analysis. In order to obtain highly purified DNA free of any inhibitory contaminant, adequate extraction methods should be applied depending on the type of initial matrix. DNA extraction requires lysing cells, inactivating cell nucleases and separating the nucleic acids from the cell debris.

2.3.1.2. *PCR-based detection method: quantitative real-time PCR*

PCR techniques can be employed after DNA isolation and purification. Classical PCR is the most commonly used nucleic acid amplification method for detecting pathogenic microorganisms. Over the last two decades, several different advances in the original PCR procedure have been described, such as multiplex PCR that allows the simultaneous detection of multiple pathogens in a food sample. The specificity of detection of this method depends on the specific binding of the primer pairs to the target sequences of the microorganisms.

However, despite being rapid, specific and sensitive, conventional PCR-based detection methods used alone do not distinguish between living and dead cells. To overcome this limitation, some DNA amplification methods – called viability PCR – integrate the use of cell viability dyes like ethidium monoazide (EMA) or propidium monoazide (PMA) (Nogva *et al.* 2003; Rudi *et al.* 2005). The criterion for determining viability is membrane integrity. In such techniques, cells are stained with EMA or PMA before DNA amplification, and the dye only binds the DNA of perforated cell membranes. Subsequent exposure of the cells to light results in the irreversible damage of nucleic acid, leading to a strong inhibition of PCR amplification. Consequently, only DNA from "living" cells, that is, with an intact membrane, is amplified (Nocker and Camper 2009; Trevors 2012; Emerson *et al.* 2017). The use of viability PCR tests for the rapid detection of foodborne pathogens has been extensively explored; among such methods, quantitative real-time PCR (qPCR) and loop-mediated isothermal amplification (LAMP) have been successfully applied (Law *et al.* 2015).

The quantitative PCR principle is based on the detection and quantification of fluorescence levels emitted during amplification, which are in turn proportional to the quantity of the accumulated PCR product. The fluorescent molecule can be either a target-specific probe labeled with a fluorescent dye together with a quencher molecule or a non-specific DNA-binding dye.

The amplification process is recorded cycle after cycle "in real time". It is then possible to draw a curve characterizing the three phases of the PCR kinetics (Figure 2.3): i) the initiation phase; ii) the exponential phase, observed when a sufficient number of PCR products have been accumulated during the initiation phase (Figure 2.3); and iii) the plateau phase, reached when one of the constituents necessary for the PCR becomes limiting. The starting point of the exponential phase is called the cycle threshold or Ct (Figure 2.3). The Ct corresponds to the number of cycles to be reached to observe a fluorescent signal significantly higher than the background noise (Figure 2.3). The Ct is inversely proportional to the logarithm of the number of DNA molecules initially present in the sample.

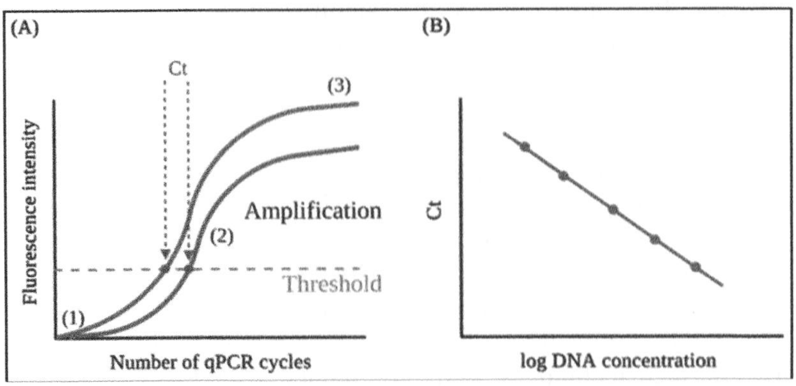

Figure 2.3. *Quantitative real-time PCR curves. (A) Real-time PCR kinetics, with (1) the initiation phase, (2) the exponential phase and (3) the plateau phase. During each cycle, the fluorescence intensity is proportional to the concentration of amplicons. The cycle threshold (Ct) is the number of cycles required for the fluorescence signal to be significantly higher than the baseline. (B) qPCR standard curve. The cycle threshold is linear with the log of the number of initial copies of target DNA. Adapted from Vidic and Auger (2019). Created with BioRender.com*

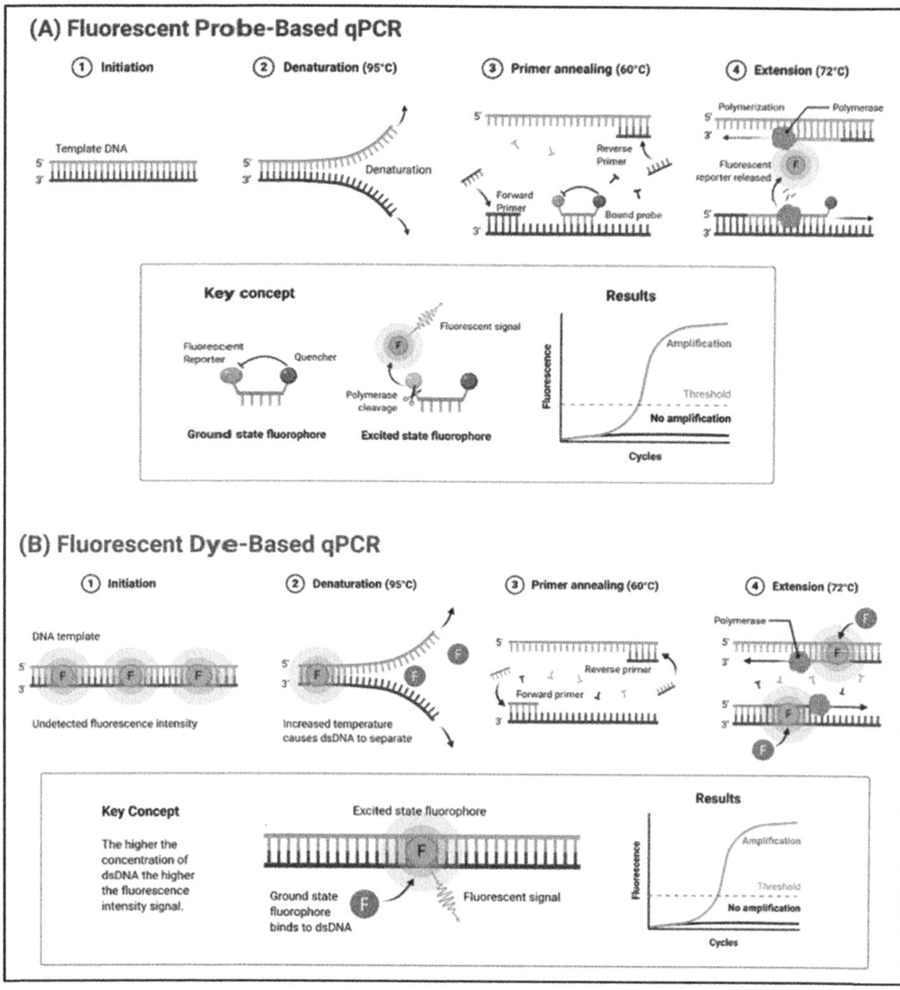

Figure 2.4. *Real-time PCR detection systems. (A) Fluorescent probe-based qPCR. The probe and primers hybridize to their target. Elongation of the complementary strand causes degradation of the probe. The fluorophore then detaches from the quencher, hence the emission of a fluorescence signal. (B) Fluorescent dye-based qPCR. When DNA is denaturated to single-stranded DNA, the fluorescent dye produces little fluorescence. After pairing of the primer to the single-stranded DNA, the dye molecules bind to the nascent double-stranded DNA, resulting in fluorescence emission upon excitation. Adapted from Fluorescent Probe-Based Real Time PCR and Fluorescent Dye-Based Real Time PCR by BioRender.com (2021). Retrieved from https://app.biorender.com/biorender-templates. For a color version of this figure, see www.iste.co.uk/membre/microbiological.zip*

Standard curves can be established to correlate the Ct values of the PCR to gene copies in the DNA extract (Figure 2.3). However, the standard curve is often obtained from a serial dilution of purified DNA, so that it may be affected by the sample composition. The efficiency or sensitivity of the PCR amplification can indeed decrease due to the turnover of interfering compounds or co-isolated DNA of a competitive flora present in the sample. Therefore, establishing calibration curves in similar environmental conditions as those of the investigated sample seems to be a prerequisite.

Two general principles exist for the quantitative detection of amplicons by qPCR: using agents that bind to double-stranded DNA (e.g. SYBR® Green) or using fluorescent probes that specifically hybridize to the amplified target fragment (e.g. Taq Man) (Figure 2.4).

qPCR techniques have been used for the detection of pathogenic bacteria (Guy *et al.* 2006) such as *L. monocytogenes* (Pan and Breidt 2007), *Campylobacter* spp. (Duarte *et al.* 2015), *E. coli* O157:H7 (Dinu and Bach 2011) and *Salmonella* (Wang *et al.* 2015). On the other hand, this method does not seem adequate for *Salmonella* spp., and enrichment followed by analysis by qPCR is more recommended (Barbau-Piednoir *et al.* 2014; Wang *et al.* 2015). In addition, a TaqMan® multiplex qPCR assay has been developed to simultaneously detect and quantify 12 common pathogens in food and drinking water: *E. coli* O157:H7, *L. monocytogenes*, *S. enterica*, *Vibrio parahaemolyticus*, β-*Streptococcus hemolyticus*, *Yersinia enterocolitica*, *Enterococcus faecalis*, *Shigella* spp., *Proteus mirabilis*, *Vibrio fluvialis*, *S. aureus* and *Campylobacter jejuni* (Liu *et al.* 2019). In this test, the detection limit for these bacteria varies between 56 and 960 CFUs/mL (Liu *et al.* 2019).

The qPCR method is highly sensitive, and can detect trace amounts of target DNA. This technique can also be automated and can quantify bacterial loads without any post-PCR handling. In addition, the use of PMA followed by qPCR is promising but requires optimization of the protocol according to the targeted bacterial species and the food matrix category.

However, the limit of quantification is around 10^3–10^4 cells per gram, which remains too high for practical application, since most samples taken along the food chain are contaminated by lower numbers of pathogens (usually less than 100 CFUs/g). Additionally, qPCR is highly dependent on primer concentration and design, and it requires stringent quality controls. A limitation of the viability of the PCR approach is that cell membrane integrity is not always a reliable indicator of cell viability. Some cells can indeed remain intact although they do not have any metabolic activity, leading to false-positive results (Ayrapetyan and Oliver 2016). Furthermore, bacterial cells can have a perforated membrane at some point during their growth, or during cell wall synthesis, so that inhibition of gene amplification can result in false-negative results (Stiefel *et al.* 2015).

2.3.1.3. *PCR-based detection method: digital droplet PCR*

In the face of certain disadvantages of qPCR, digital droplet PCR (ddPCR) is an end-point measurement that overcomes the dependency on a DNA calibrant. This technology allows the absolute quantification of a target DNA without relying on the measurement of the efficiency of the PCR reaction or on Ct determination. In addition, the DNA amplification reaction is less sensitive to the presence of inhibitors, possibly as a result of dilution and compartmentalization. Finally, it does not require a standard curve. ddPCR is more expensive than qPCR but its high sensitivity and high precision make it a method of choice for detecting and quantifying low copy numbers of target DNA.

The digital droplet PCR is based on dividing the target DNA molecules into a large number of oil-dispersed aqueous droplets also containing the PCR mixture in order to carry out a multitude of PCRs simultaneously. Each droplet constitutes a separate reaction compartment and contains 0 or 1 (or more) template copies (Hindson *et al.* 2011). After amplification by a conventional PCR, a Poisson distribution is used to extrapolate the fraction of positive end-point reactions. The Poisson distribution is a function that relates the probability of a given number of events occurring independently in a given sample when the average rate of occurrence is known.

Different strategies have been developed to produce the microdroplets to capture the amplified DNAs and to handle them after the PCR. Microfluidics is one strategy; it requires the handling of very small quantities of fluids (from 10^{-3} to 10^{-12} μL) in microchannels (Vidic and Auger 2019). By using two-phase flows (a combination of a continuous oil phase and a discontinuous aqueous phase), microfluidic systems produce monodispersed micrometric drops at high flow rates (Vidic and Auger 2019). In some of the marketed microfluidic systems – for example, the QX100/200 system (Bio-Rad) – encapsulation of the DNA molecules in 10,000–20,000 microdroplets is followed by their injection into a thermal cycler to carry out the PCR (Figure 2.5). Then, the droplets are re-injected into a reader that measures the fluorescence of each droplet at the end-point of the PCR reaction using fluorescent probes or molecules intercalating DNA (Figure 2.5). Consequently, the PCR-positive and PCR-negative droplets are individually counted on the droplet reader by passing them through a fluorescence detector in a single stream in a manner similar to flow cytometry. The number of target DNA molecules initially present can be determined from the numbers of positive and negative droplets after amplification by probability analysis and readjustment according to Poisson's law.

Digital PCR technology is booming in the field of food microbiology. It has been used to detect low concentrations of pathogens in food products (Porcellato *et al.* 2016; Peruzy *et al.* 2020) and mixtures of several food pathogens present in foodstuffs (Bian *et al.* 2015; Cremonesi *et al.* 2016). In addition, a statistically significant correlation has been observed between ddPCR and the culture-based methods (ISO standard) for *Campylobacter* quantification in poultry products (Papić *et al.* 2017). However, ddPCR overestimated the level of contamination on broiler neck-skin samples due to a high number of false-positive results and probably the very high sensitivity of detection that led to the detection of residues of dead *Campylobacter* (Papić *et al.* 2017).

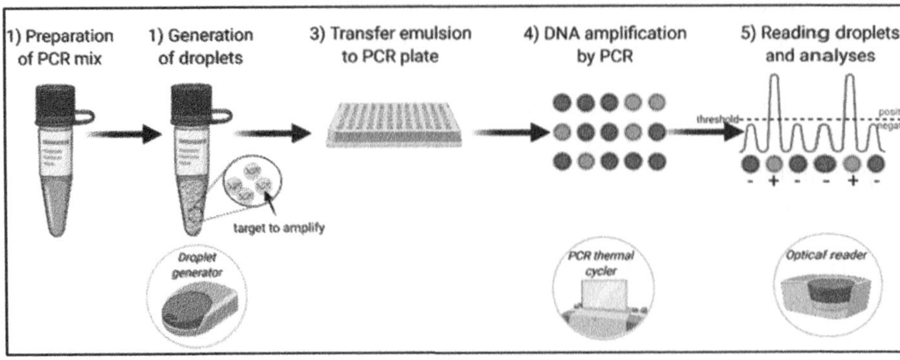

Figure 2.5. *Main steps of digital droplet PCR (ddPCR). Step 1: preparation of the PCR mix. Step 2: partitioning. DNA molecules are randomly distributed among 20,000 droplets. Step 3: transfer of the emulsion to a PCR plate. Step 4: DNA amplification. The target sequence is amplified by end-point PCR in each droplet. Step 5: reading and analysis. The numbers of positive and negative droplets are counted to give a precise quantification of the target sequences in the sample. Created with BioRender.com. For a color version of this figure, see www.iste.co.uk/membre/microbiological.zip*

2.3.2. *Molecular typing of foodborne pathogens*

Molecular subtyping methods have proven invaluable for tracking and tracing pathogens along the food chain by helping to identify sources of infection and the transmission route. Management of foodborne illness threats requires quick detection of the foodborne disease and understanding and responding to it. For this purpose, public health authorities around the world need surveillance and response systems capable of (i) early detection of food safety events and foodborne outbreaks and (ii) monitoring trends in priority foodborne diseases to assess the movement and variation of human pathogens. Surveillance of foodborne pathogens, and mainly their involvement in outbreaks, is made possible by molecular techniques like molecular typing.

The molecular typing of foodborne pathogens, such as *Salmonella*, *L. monocytogenes*, *E. coli* EHEC/STEC and *Campylobacter*, helps us to identify the strains specific to the origin of foodborne outbreaks and detect new potential health threats. By relating strains to types of food, it is possible to assess the role that different foods play in human infections. Molecular typing of microbes has undergone rapid development in recent years.

Pulsed-field gel electrophoresis (PFGE) is a powerful molecular typing technique whereby genomic DNA is isolated from the organism of interest, followed by restriction enzyme analysis. Genomic DNA is fragmented with rare-cutting enzymes and the digestion products are analyzed on an agarose gel by applying an electric field that periodically changes direction. This way, the larger DNA fragments (entire genomic DNA) are separated, and their length is measured approximately. The resulting size and number of DNA fragments form a stable and reproducible restriction pattern, also called a DNA fingerprint, which can be compared between different strains.

Figure 2.6 illustrates the main steps of the PFGE method and shows an example of an agarose gel where each lane represents a DNA fingerprint or pattern. PFGE is different from conventional DNA electrophoresis in that it can separate very large DNA fragments (up to 10 Mb) and generate a fingerprint by constantly changing the direction of the electric field.

Several limitations of the PFGE technique have been identified. Among them we can note that it is a time-consuming technique and imperfect discrimination and technical variability make it unfit for certain strains.

One of the most successful sequence-based typing approaches is the concept of multilocus sequence typing (MLST). In general, MLST involves the amplification of seven loci of housekeeping genes by PCR, followed by DNA sequencing of the resulting PCR fragments. Specific DNA sequences are then matched to allelic profiles. A single nucleotide variation at any of these loci defines a different allele and a sequence type (ST).

The discriminatory power of MLST is generally comparable to or slightly better than traditional serotyping (Achtman *et al.* 2012). Nevertheless, 7-gene MLST is often not discriminative enough for outbreak detection. As a result, the PCR-based multilocus variable-number tandem-repeat (MLVA) method was developed to discriminate between highly related strains (Lindstedt 2005). This approach is based on the detection of repetitive tandem DNA units within various loci. Repeating units are approximately 1–100 base pairs in length. The number of tandem repeats can change due to slipped strand mispairing with each generation, so that the relatedness of bacteria can be inferred from tandem repeat unit variation. As MLVA has been proven to be a fast-tracing tool with better resolution than PFGE in outbreak studies, the method has been standardized for certain pathogenic subtypes (Nadon *et al.* 2013).

2.3.3. *Next-generation sequencing-based methods*

The next-generation sequencing (NGS) approach leads to the sequencing of an entire bacterial genome in small random fragments (<100 to several thousand bp) multiple times in a single reaction. This step is followed by the determination or reconstitution of the full DNA sequence electronically by linking fragments with overlapping sequences using sequence assembly software. NGS in food microbiology is predominantly used in two ways: (i) to determine the whole genome sequence of a single cultured isolate, commonly referred to as "whole genome sequencing" (WGS), and (ii) in "metagenomics", where NGS is applied to a biological sample generating sequences of multiple (if not all) microorganisms in that sample.

The high discriminatory power of WGS compared to traditional molecular typing tools is well-established, and WGS is gaining acceptance as a prospective surveillance tool for foodborne illnesses. The standard typing methods for many foodborne bacterial pathogens (e.g. antigen testing, PFGE or MLVA) have shown a lower level of strain discrimination and are not informative enough for certain types of analysis, such as evolutionary studies and spatiotemporal investigations. In contrast, WGS provides the highest possible microbial subtyping resolution to public health authorities for surveillance of and response to foodborne illnesses. This approach also offers genomic information for the prediction of phenotypes (serotyping, antimicrobial resistance, biofilm formation, pathogenicity and virulence). The main steps of WGS are described in Figure 2.6.

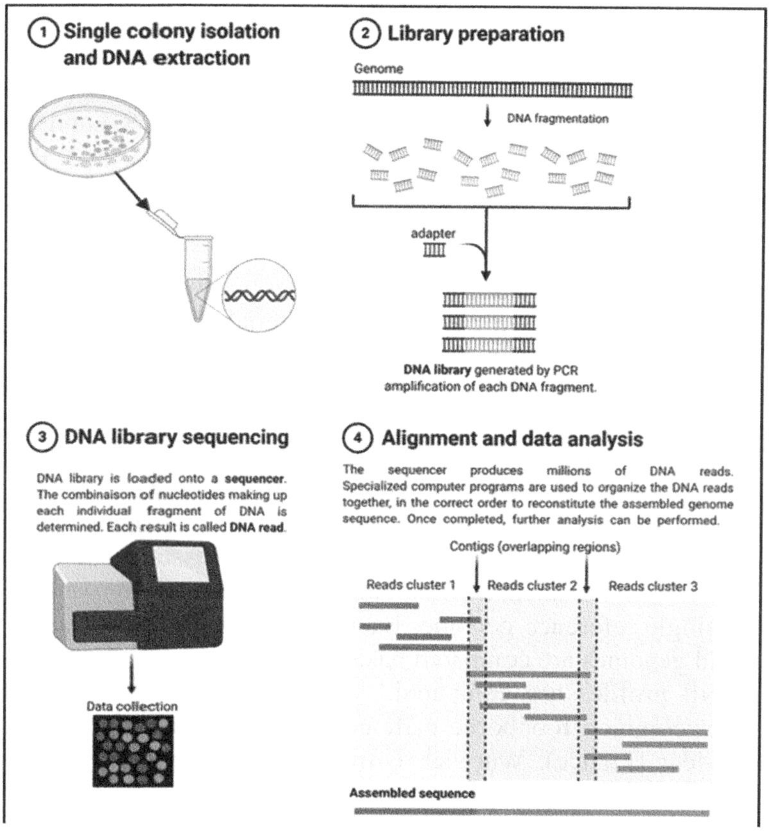

Figure 2.6. *Main steps of whole genome sequencing (WGS). Adapted from Next Generation Sequencing (Illumina) by BioRender.com (2021). Retrieved from https://app.biorender.com/biorender-templates. For a color version of this figure, see www.iste.co.uk/membre/microbiological.zip*

Two analytical post-sequencing approaches are available to analyze genomic data and determine the relatedness between bacterial isolates: base-by-base (single-nucleotide polymorphism (SNP)) or gene-by-gene (MLST) analysis. Both approaches are used to link clinical isolates back to environmental or contaminated food sources, or to cluster food and environmental isolates across time and space for the detection of outbreaks.

In the SNP-based approach, the sequencing reads of the tested strain are aligned or mapped to a known sequenced reference genome to determine the nucleotide differences at specific positions in both coding and non-coding regions (Davis *et al.* 2015). For each isolate, every SNP relative to the reference genome is recorded and then used to quantify the genetic relatedness between strains. Several validated SNP-based tools can be found in the public domain, such as those developed by the US Food and Drug Administration (FDA), the Center for Disease Control and Prevention (CDC), Public Health England (PHE) and the Center for Genomic Epidemiology (CGE) (Jagadeesan *et al.* 2019).

The MLST-based approach consists of assessing variation in the coding regions, that is, the genes (or loci) of a bacterial genome (Maiden *et al.* 2013). Three levels of discrimination can be reached based on the number of genes: (i) seven-housekeeping-gene (7-gene) MLST; (ii) core genome (cg) MLST, which assesses genes present in almost all strains of a given species; and (iii) whole-genome (wg) MLST, which analyzes almost all the genes present in any strain of a given species. The genes identified in the strains are compared against a reference database of genes with all known gene variants ("alleles") of the species/pathogen to be assessed. The reference in MLST is therefore a database of loci and alleles from numerous strains and not just a single reference genome. Each unique allele sequence is given a number, and genomes are compared based on the allele numbers, in the same way as SNP profiles are compared. Several publicly available cgMLST schemes for common foodborne pathogens are available and hosted by the Institut Pasteur (France), Warwick University (UK) and the University of Oxford (UK) (Jagadeesan *et al.* 2019).

WGS is increasingly used in public health and food safety laboratories. However, there is great variation in the level of implementation of WGS in laboratories in different sectors and in different countries and with regard to the type of analysis and data interpretation.

According to the EFSA *et al.* (2018), WGS was already used in the reporting laboratories of 17 of the 30 European Union and associated countries in late 2016. In particular, WGS was found to be used in all the EU reference laboratories consulted, almost half of the national reference laboratories (44% of affirmative respondents) and only a few official laboratories (7% of affirmative respondents) (EFSA *et al.* 2018). The main reason for not implementing WGS was the lack of capacity (budget and skills) (EFSA *et al.* 2018). WGS has been used primarily for outbreak investigation, surveillance monitoring and for the analysis of *L. monocytogenes*, *E. coli* and *Salmonella.* A significant number of foodborne illness outbreaks have been detected and controlled since 2011 using WGS. For example, during the 2011 European outbreak caused by Shiga toxin-producing *E. coli* (STEC), which caused 4,000 cases and more than 50 deaths, molecular typing played a key role in identifying cases and demonstrated that a single strain was responsible for the epidemic outbreaks in Germany and France. WGS has been used for routine monitoring of *L. monocytogenes* since March 2017 and is being applied to support foodborne outbreak investigations caused by a large number of other pathogens in the United States (Jackson *et al.* 2016). One of the largest applications of WGS in foodborne disease surveillance was the *L. monocytogenes* outbreak in South Africa in 2017–2018.

WGS has become a powerful tool in food safety and may play a pivotal role in food safety surveillance and risk assessment. The overall costs of sequencing have gone down to almost similar levels as the price range of traditional methods used to detect pathogens in food products or food environments. In the context of the environmental monitoring of food production, WSG can be used as a powerful pathogen-tracking tool to trace the sources of contamination. In addition, according to Brown *et al.* (2019), a few companies are already using WGS to identify contaminated niche locations along their production lines and prevent their end-products from ever being contaminated (Brown *et al.* 2019).

Several barriers still need to be lifted before these techniques can be incorporated in foodborne pathogen detection. Genomic technologies are not always suitable for addressing two fundamental challenges in the detection and enumeration of pathogens, that is, sensitivity and the determination of viability. Foodborne pathogens are usually found in very low numbers, and this creates specific limitations for epidemiological studies.

Limitations in the implementation of the WGS approach in the industry concern the need for highly trained bioinformatics professionals that smaller companies cannot usually afford and the uncertainty as to the frequency with which this technology will be applicable in their own microbiological control systems. Furthermore, it seems necessary to standardize and validate protocols and to develop user-friendly and powerful computer platforms to process and facilitate the interpretation of the massive databases created by the method.

Other NGS-based methods that do not require bacterial culture can be applied to monitor the diversity of microbial communities in food or food processing lines (Figure 2.7). Amplicon sequencing or metabarcoding involves the amplification and sequencing of specific marker genes, such as the 16S rRNA gene. Metagenomics consists of the random shotgun sequencing of the whole genomic content of communities. Metagenomic tools can help anticipate microbial spoilage by studying changes in the diversity or proportion of spoilage-associated microbes in the microbiota of food products (Ercolini *et al.* 2011; Kable *et al.* 2016). Although pathogens are present at a low level in foodstuffs, these techniques can be used to explore the detection, identification and characterization of pathogens in food (Leonard *et al.* 2015, 2016; Aw *et al.* 2016) and food chain environments (Yang *et al.* 2016). However, this investigation requires several hours of culture-based enrichment coupled with high-sequencing analysis.

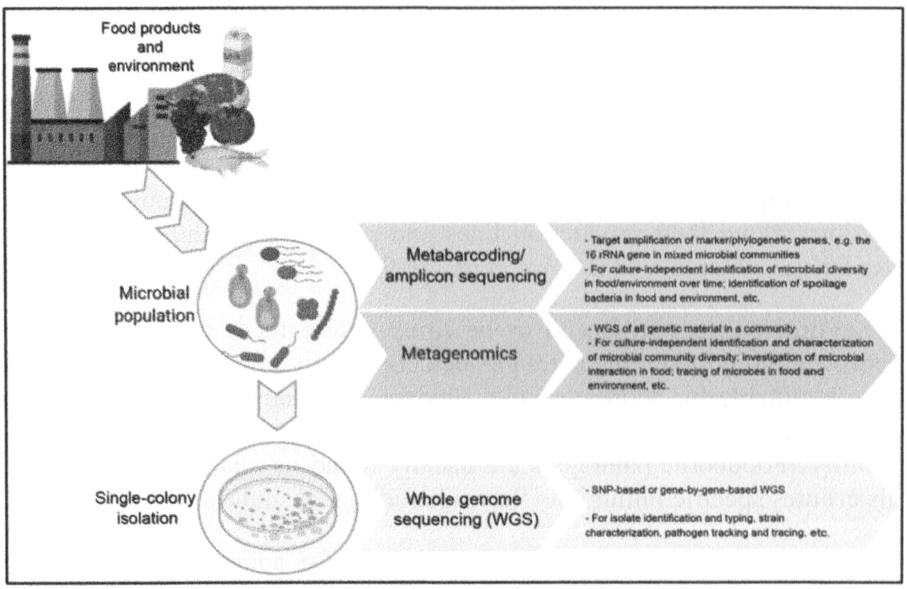

Figure 2.7. *Summary of the potential use of NGS by the food industry according to Jagadeesan* et al. *(2019). Created with BioRender.com*

2.4. Immunology-based assays

Detection of foodborne pathogens by immunological methods is based on antibody–antigen interactions, in which a particular antibody binds to its specific antigen. The sensitivity and specificity of immunological methods are dependent on the binding strength of the particular antibody with its antigen. Immunological methods involve the use of polyclonal and monoclonal antibodies.

2.4.1. *Enzyme-linked immunosorbent assay*

The enzyme-linked immunosorbent assay (ELISA) is one of the most commonly used immunological methods for the detection of foodborne pathogens. Different formats of ELISA are available, but almost all of them follow the same principles and key operations, as illustrated in Figure 2.9.

In a typical ELISA, the antigens are immobilized either by direct adsorption or via an antibody adsorbed at the bottom of microplate wells (Figure 2.8). The antigen is detected with a specific detection antibody that can be either (i) directly labeled with a signal-generating enzyme or fluorophore, or (ii) secondarily probed with an enzyme- or flour-labeled secondary antibody (or avidin–biotin chemistry). For enzymatic detection, the appropriate enzyme substrate is added to the wells. The positive or negative result can be observed visually, or absorbance can be measured. The signal is proportional to the amount of antigen in the sample. Washing between each step ensures that only specific (high-affinity) binding is maintained to produce the signal at the final step. ELISA principles, methods and applications have been reviewed in numerous publications (Crowther 2009; Hnasko 2015).

The sandwich ELISA is the most effective ELISA and involves two antibodies (Figure 2.8). The primary antibody is usually immobilized at the bottom of the assay plate. The target antigen – for example, bacterial cells or bacterial toxins from the food sample – binds to the immobilized primary antibody. After that, an enzyme-conjugated secondary antibody is added and binds to the antigen. Intermediate washes allow the removal of unbound antigens or antibodies. The complex made of the antigen sandwiched between two antibodies is formed and it can be detected by adding a colorless substrate that is converted into a colored form in the presence of the enzyme (Figure 2.8). Different types of enzymes can be used in ELISA, including horseradish peroxidase (HRP), alkaline phosphatase and beta-galactosidase.

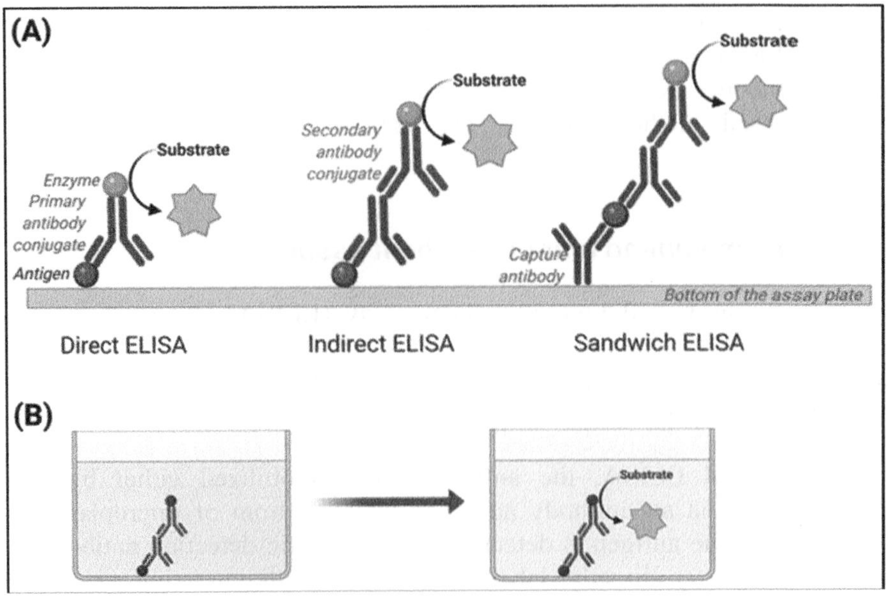

Figure 2.8. *Overview of common enzyme-linked immunosorbent assay (ELISA) formats. (A) Variants of ELISA. The antigen of interest is immobilized either by the direct adsorption to the assay plate (direct and indirect ELISA) or indirectly via a capture antibody attached to the plate (sandwich ELISA). The antigen is then detected either directly (using a labeled primary antibody in the direct ELISA) or indirectly (using a labeled secondary antibody in the indirect or sandwich ELISA). Detection is performed by measuring the activity of the reporter enzyme via incubation with the appropriate substrate to produce a measurable product. (B) Color changes occur upon substrate binding. Adapted from ELISA Overview by BioRender.com (2021). Retrieved from https://app.biorender.com/biorender-templates. For a color version of this figure, see www.iste.co.uk/membre/ microbiological.zip*

Several studies have been performed using an ELISA for the rapid detection of foodborne pathogens or toxins. For example, Kumar *et al.* (2011) detected pathogenic *V. parahaemolyticus* in seafood with a sandwich ELISA, using monoclonal antibodies against the TDH-related hemolysin (TRH) produced by the pathogen (Kumar *et al.* 2011). The detection limit of this assay was 10^3 cells of pathogenic *V. parahaemolyticus*. ELISA is also commonly used for the detection of toxins present in foods such as

Clostridium perfringens α, β and ε toxins, staphylococcal enterotoxins A, B, C and E, botulinum toxins and *E. coli* enterotoxins (Aschfalk and Müller 2002; Zhao *et al.* 2014). Several validated commercial ELISA systems designed for the rapid detection of *Salmonella* in raw or processed food products are available: Assurance GDS™ for *Salmonella* (BioControl Systems, Inc., Bellevue, WA), TECRA *Salmonella* (Tecra International Pty Ltd, French Forest, New South Wales, Australia) and *Salmonella* ELISA Test SELECTA/OPTIMA (Bioline APS, Denmark) (Bolton *et al.* 2009).

High-throughput ELISA systems such as VIDAS® are advanced testing solutions for pathogen detection in food products. The VITEK immunodiagnostic assay system (VIDAS) performs the entire ELISA procedure automatically. It utilizes enzyme-linked fluorescent immunoassay (ELFA), which is similar to ELISA but more sensitive in reporting the results (Fung 2002). Used by thousands of laboratories worldwide, the VIDAS® solution offers more than 40 international validations (including ISO 16140, AOAC RI, AOAC OMA and more). The automated food pathogen detection system is optimized for food production and farm applications.

As for a typical ELISA, VIDAS has also been applied to detect several foodborne pathogens in different food products, such as *Salmonella* (Vieira-Pinto *et al.* 2007; Gómez-Govea *et al.* 2012), *L. monocytogenes* (Vaz-Velho *et al.* 2000; Gómez-Govea *et al.* 2012) and *E. coli* STEC (Carvalho *et al.* 2014). In addition, immune-based assays have been so far the major choice for the detection of staphylococcal enterotoxins in food. Since as little as 100–200 ng of the toxin can cause symptoms of staphylococcal intoxication, the methods have to be very sensitive (Evenson *et al.* 1988; Boyle *et al.* 2010). The sensitivity of VIDAS SET2 can be lower than 1 ng/g of different staphylococcal toxins, with 100% specificity (Vernozy-Rozand *et al.* 2004).

2.4.2. *Latex agglutination assay*

The other immune-based system commonly used to detect pathogens or toxins in food is the reversed passive latex agglutination (RPLA) assay. The agglutination technique employs latex beads coated with specific antibodies or normal serum which react with antigens on the surface of bacterial cells to form visible aggregates (Mahon and Lehman 2018). The difference in the reaction with control and sensitized latex is read visually.

The assays are specific, simple and reliable, so they have generally been used as a confirmatory analysis technique rather than as a screening test for *Salmonella* strains (Eijkelkamp *et al.* 2009; Love and Sobsey 2007). Several commercial kits are available on the market, including Spectate (May & Baker Diagnostics Ltd., Glasgow, Scotland, UK), Wellcolex color *Salmonella* (Wellcolex, Merseyside, UK), *Salmonella* Latex test (Oxoid, Basingstoke, UK), Bactigen (Wampole Laboratories, Cranbury, NJ, USA) and Slidex (BioMérieux, Marcy L'Etoile, France).

Immune-based assays are comparable to and often more rapid and specific than cultural methods (Keith 1997; Uyttendaele *et al.* 2003). These techniques can be automated to handle large samples and to reduce time and labor. As routine analysis techniques, they remain a valuable option because they can detect viable non-culturable cells (Maciorowski *et al.* 2006).

However, these methods present potential shortcomings for the detection of pathogens in food, including cross-reactions with closely related antigens, antigen variation and high automation costs. It is also important to understand that a longer enrichment time is often needed to obtain the appropriate number of cells, and the sensitivity of the detection can be limited due to the food matrices and stressed cells.

2.5. Other methods

2.5.1. *Biosensor-based methods*

A biosensor is an analytical device composed of two main elements:

– a bioreceptor that recognizes the target analyte;

– a transducer that converts the biological interactions into a measurable electrical signal.

The bioreceptor can be (i) biological material like enzymes, antibodies, nucleic acids and cell receptors, (ii) biologically derived material like aptamers and recombinant antibodies or (iii) biomimetic material like imprinted polymers and synthetic catalysts. This material is immobilized on the surface of a physicochemical transducer or transducing microsystem.

Enzyme biosensors are highly selective, rapid and easy to use, but they are expensive and sometimes lose their activity when immobilized on a transducer. An enzyme biosensor consists of an enzyme such as glucose oxidase, galactosidase, lactate oxidase, etc. (Fung 2002), with a transducer to produce a signal resulting from different physicochemical changes caused by enzyme-catalyzed reactions. The signal is converted into a measurable response (an electric current, a temperature change, light absorption) to quantify the target analyte concentration (Baeumner 2003). In addition, antibodies are the most commonly used biological recognition element due to their high specificity and affinity to antigens, their diversity and their availability on the market. Antibody-based biosensors integrate an antibody as a biological recognition element attached to a signal transducer to determine the target analyte concentration (Rogers 2000). Antibody-based biosensors can be divided into three main assay formats, the characteristics of which are largely determined by the type of signal transduction: immobilized antibodies, immobilized antigens and non-immobilized antibodies or antigens (Rogers 2000).

The signal recognition and transduction in the biosensor are achieved using different types of transducers: optical, electrochemical, mass-based, thermometric, micromechanical or magnetic (Velusamy *et al.* 2010; Zhao *et al.* 2014). A recognition signal is generated when a specific analyte binds to the biological recognition element. The signal can be a change in mass, oxygen consumption, potential difference, refractive index, pH, current or other parameter (Baeumner 2003; Eijkelkamp *et al.* 2009).

The commercially available biosensors for the detection of foodborne pathogens are mostly optical-based biosensors. Commercial biosensors offer varying degrees of automation (Leonard *et al.* 2003). Due to its sensitivity, the surface plasmon resonance (SPR) biosensor is the most commonly used optical biosensor for the detection of foodborne pathogens. SPR employs reflectance spectroscopy for pathogen detection (Velusamy *et al.* 2010). In SPR, bioreceptors are immobilized on the surface of a thin metal. The electromagnetic radiation of a certain wavelength interacts with the electron cloud of the thin metal and produces a strong resonance. When the pathogen binds to the metal surface, this interaction alters its refractive index, which modifies the wavelength required for electron resonance (Zhang 2013; Zhao *et al.* 2014).

Commercial optical biosensors using SPR techniques such as the SPREETA biosensor and BIACORE 3000 biosensor are currently available for the detection of foodborne pathogens. The SPREETA biosensor was successfully used to detect *E. coli* O157:H7 in milk, apple juice and ground beef with a limit of detection of around 10^2–10^3 CFUs/mL (Waswa *et al.* 2007), and also to detect *Salmonella* enteritidis and *Salmonella* typhimurium

(Son *et al.* 2007; Lan *et al.* 2008). In addition, *L. monocytogenes* was successfully detected with a BIACORE 3000 biosensor with a detection limit of 10^5 cells/mL (Leonard *et al.* 2004).

Electrochemical biosensors are categorized into several types – amperometric, impedimetric, potentiometric and conductometric – according to their measurement of changes caused by the antigen–bioreceptor interactions in current, impedance, voltage and conductance, respectively (Zhang 2013). This technique was successfully used to detect foodborne pathogens like *Bacillus cereus* in fruit, vegetables, fried rice and cooked corn, with a limit of detection of around 35–88 CFUs/mL (Pal *et al.* 2008). An amperometric magnetoimmunosensor was employed for the detection of *S. aureus* with a limit of detection of 1 CFU/mL, and the analysis time was 2 h (Esteban-Fernández de Ávila *et al.* 2012). An antibody-based potentiometric biosensor allowed the detection of *E. coli* in vegetable food with a detection limit of 10 cells/mL (Ercole *et al.* 2003), whereas impedimetric spectroscopy allowed for the detection and quantification of *E. coli* with a more variable detection limit (Muñoz-Berbel *et al.* 2008).

A recent review published by Ali *et al.* (2020) presents the effectiveness of various sensing technologies for the detection of foodborne pathogens in food products, as well as the criteria for industrial use of this technology (Ali *et al.* 2020).

Biosensors do not require sample pre-enrichment, unlike nucleic acid-based methods and immunological methods, which require sample pre-enrichment to concentrate the pathogens before detection (Singh *et al.* 2013). However, biosensors are commercialized more slowly than other rapid methods because of several factors, for example, cost, quality assurance, stability issues, sensitivity issues and instrumentation design. Producing inexpensive and reliable sensors is difficult because storing, stabilizing and calibrating biosensors and fully integrating the sensor system are tricky (Velasco-Garcia and Mottram 2003).

2.5.2. *MALDI-TOF mass spectrometry*

Matrix-assisted laser desorption/ionization time-of-flight (MALDI-TOF) mass spectrometry can be used to confirm and identify foodborne pathogens present in food matrices. It is faster than traditional microbiology methods and has the same level of discrimination; it is cheaper and requires less technical expertise than genotypic methods. It is used in analytical laboratories for routine food microbiology where time plays a key role because these laboratories primarily work with the food industry.

MALDI-TOF mass spectrometry applied to microbiology can identify microorganisms down to the species level. The principle is based on the ionization of bacterial proteins by a laser beam and the creation of characteristic peaks (the spectrum). This characteristic spectrum, called the peptide mass fingerprint, is specific to each microorganism. The microorganism is identified by comparing its fingerprint profile with a validated database of spectra. The associated software program searches for similarity to the bacterial species according to an index of reliability between the two spectra. However, MALDI-TOF mass spectrometry does not provide information on the serotype or the pathogenicity of the species. A mass spectrometer is typically made up of three elements: an ion source (MALDI), a system that separates molecules (TOF) and a detection device.

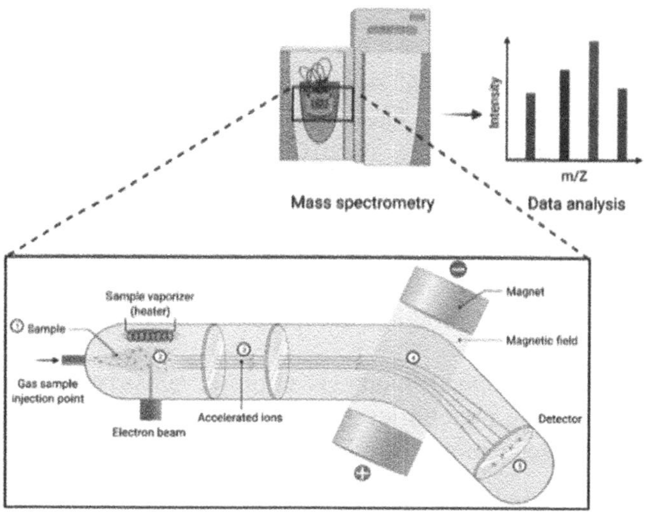

Figure 2.9. *Matrix-assisted laser desorption/ionization time-of-flight (MALDI-TOF) mass spectrometer principle. A sample is injected into the mass spectrometer (1), and the molecules are ionized (2) and accelerated (3). The ions are then separated by mass and charge by the mass analyzer via electromagnetic deflection (4), and the ions that are properly aligned are detected and amplified (5). The entire system is under intense vacuum during the entire process. After signal amplification, the data that are generated reports the relative abundance of each ion based on its mass-to-charge (m/z) ratio. Adapted from Mass Spectrometry Principle by BioRender.com (2021). Retrieved from https://app.biorender.com/biorender-templates. For a color version of this figure, see www.iste.co.uk/membre/microbiological.zip*

Figure 2.9 summarizes the principle of a MALDI-TOF mass spectrometer. The analyte is first co-crystallized with small organic compounds (a matrix) to protect it from direct contact with the ionizing beam and to prevent degradation. These absorb radiation from the UV laser and transfer energy to proteins that ionize positively. The generated ions are released from the proteins. Then, the charged molecules (+1) are accelerated in an electric field.

After their passage through the electric field, the ions enter a tube not subjected to an electric field and are separated according to their mass charge. Ions take on different speeds because large molecules move more slowly than small ones. Then, a time-of-flight analyzer detects the passage of each ion; this creates the peaks on the spectrum, which is compared with a database. Some components of the spectrum are specific to a genus, others to a species or even a subspecies. Depending on the quality and purity of the sample and the number of reference spectra in the database, bacterial identification can be done within seconds up to a few minutes.

Preparation of the analytical sample is a critical step in MALDI-TOF mass spectrometry analyses. Theoretically speaking, the preparation of bacterial, yeast and mold samples is universal. In order to obtain reliable and reproducible identification results, bacterial growth conditions (incubation time and temperature, culture medium) have to be similar across tests. The analysis starts from a pure bacterial culture. Analyses should be performed with fresh bacterial cultures because cold stress or nutrient deficiency can change the composition of proteins and therefore affect the results.

MALDI-TOF mass spectrometry is a rapid, sensitive and inexpensive technique for analyzing intact microbial cells or cell extracts. This technique has several advantages, among which:

– results are obtained quickly;

– it discriminates between more than 2,500 species of microorganisms (lactic acid bacteria, propionic bacteria, micrococci, non-pathogenic staphylococci, yeasts and molds), spoilage (psychotrophic flora, species belonging to *Enterobacteriaceae*, thermoresistant flora, *Bacillus*, *Staphylococci*, *Clostridia*, yeasts and molds) and pathogens;

– it can be applied to all strains derived from food matrices and production environments.

However, some limitations of MALDI-TOF mass spectrometry for pathogen identification have been described. In particular, the comparison of the fingerprint of the unknown isolate to reference fingerprints from the database is the most crucial step for species identification. Consequently, the database needs to contain not only reference fingerprints of all the species of interest, but also fingerprints of multiple strains of each species (Lartigue *et al.* 2009). In addition, MALDI-TOF mass spectrometry has been found to fail to distinguish species that are phylogenetically very close, such as *E. coli* and *Shigella* spp. (Bizzini *et al.* 2010).

2.6. References

Achtman, M., Wain, J., Weill, F.X., Nair, S., Zhou, Z., Sangal, V., Krauland, M.G., Hale, J.L., Harbottle, H., Uesbeck, A., Dougan, G., Harrison, L.H., Brisse, S. (2012). Multilocus sequence typing as a replacement for serotyping in *Salmonella enterica*. *PLoS Pathogens*, 8(6), e1002776.

Ali, A.A., Altemimi, A.B., Alhelfi, N., Ibrahim, S.A. (2020). Application of biosensors for detection of pathogenic food bacteria: A review. *Biosensors*, 10(6), 58.

Aschfalk, A. and Müller, W. (2002). *Clostridium perfringens* toxin types from wild-caught Atlantic cod (*Gadus morhua L.*), determined by PCR and ELISA. *Canadian Journal of Microbiology*, 48(4), 365–368.

Aw, T.G., Wengert, S., Rose, J.B. (2016). Metagenomic analysis of viruses associated with field-grown and retail lettuce identifies human and animal viruses. *International Journal of Food Microbiology*, 223, 50–56.

Ayrapetyan, M. and Oliver, J.D. (2016). The viable but non-culturable state and its relevance in food safety. *Current Opinion in Food Science*, 8, 127–133.

Baeumner, A.J. (2003). Biosensors for environmental pollutants and food contaminants. *Analytical and Bioanalytical Chemistry*, 377(3), 434–445.

Barbau-Piednoir, E., Mahillon, J., Pillyser, J., Coucke, W., Roosens, N.H., Botteldoorn, N. (2014). Evaluation of viability-qPCR detection system on viable and dead *Salmonella* serovar enteritidis. *Journal of Microbiological Methods*, 103, 131–137.

Bell, C., Neaves, P., Williams, A.P. (2005). *Food Microbiology and Laboratory Practice*. Blackwell Publishing, Iowa.

Bhunia, A.K. (2014). One day to one hour: How quickly can foodborne pathogens be detected? *Future Microbiology*, 9(8), 935–946.

Bian, X., Jing, F., Li, G., Fan, X., Jia, C., Zhou, H., Jin, Q., Zhao, J. (2015). A microfluidic droplet digital PCR for simultaneous detection of pathogenic *Escherichia coli* O157 and *Listeria monocytogenes*. *Biosensors and Bioelectronics*, 74, 770–777.

Bird, P., Bastin, B., Klass, N., Crowley, E., Agin, J., Goins, D., Bakken, H., Lingle, C., Schumacher, A. (2020). Evaluation of the 3M™ Petrifilm™ Rapid *E. coli*/coliform count plate for the enumeration of *E. coli* and coliforms: Collaborative study, first action: 2018.13. *Journal of AOAC International*, 103(2), 513–522.

Bizzini, A., Durussel, C., Bille, J., Greub, G., Prod'hom, G. (2010). Performance of matrix-assisted laser desorption ionization-time of flight mass spectrometry for identification of bacterial strains routinely isolated in a clinical microbiology laboratory. *Journal of Clinical Microbiology*, 48(5), 1549–1554.

Bolton, F.J., Fritz, E., Poynton, S., Jensen, T. (2009). Rapid enzyme-linked immunoassay for detection of *Salmonella* in food and feed products: Performance testing program. *Journal of AOAC International*, 83(2), 299–304.

Boyle, T., Njoroge, J.M., Jones, R.L.J., Principato, M. (2010). Detection of staphylococcal enterotoxin B in milk and milk products using immunodiagnostic lateral flow devices. *Journal of AOAC International*, 93(2), 569–575.

Brown, E., Dessai, U., McGarry, S., Gerner-Smidt, P. (2019). Use of whole-genome sequencing for food safety and public health in the United States. *Foodborne Pathogens and Disease*, 16(7), 441–450.

Carvalho, R.N., De Oliveira, A.N., De Mesquita, A.J., Rezende, C.S.M., De Mesquita, A.Q., Romero, R.A.M. (2014). PCR and ELISA (VIDAS ECO O157®) *Escherichia coli* O157:H7 identification in Minas Frescal cheese commercialized in Goiânia, GO. *Brazilian Journal of Microbiology*, 45, 07–10.

Center for Disease Control (2021). Pulsed-field gel electrophoresis [Online]. Available at: https://www.cdc.gov/pulsenet/pathogens/pfge.html [Accessed 31 January 2021].

Cremonesi, P., Cortimiglia, C., Picozzi, C., Minozzi, G., Malvisi, M., Luini, M., Castiglioni, B. (2016). Development of a droplet digital polymerase chain reaction for rapid and simultaneous identification of common foodborne pathogens in soft cheese. *Frontiers in Microbiology*, 7(1725), 10.

Crowther, J.R. (2009). *The ELISA Guidebook*. Humana Press, New York.

Davis, S., Pettengill, J.B., Luo, Y., Payne, J., Shpuntoff, A., Rand, H., Strain, E. (2015). CFSAN SNP Pipeline: An automated method for constructing SNP matrices from next-generation sequence data. *PeerJ Computer Science*, 1, e20.

Dewey-Mattia, D., Manikonda, K., Hall, A.J., Wise, M.E., Crowe, S.J. (2018). Surveillance for foodborne disease outbreaks – United States, 2009–2015. *MMWR Surveillance Summaries*, 67(SS-10), 1–11.

Dinu, L.D. and Bach, S. (2011). Induction of viable but nonculturable *Escherichia coli* O157:H7 in the phyllosphere of lettuce: A food safety risk factor. *Applied and Environmental Microbiology*, 77(23), 8295–8302.

Duarte, A., Botteldoorn, N., Coucke, W., Denayer, S., Dierick, K., Uyttendaele, M. (2015). Effect of exposure to stress conditions on propidium monoazide (PMA)-qPCR based *Campylobacter* enumeration in broiler carcass rinses. *Food Microbiology*, 48, 182–190.

Dwivedi, H.P. and Jaykus, L.-A. (2011). Detection of pathogens in foods: The current state-of-the-art and future directions. *Critical Reviews in Microbiology*, 37(1), 40–63.

EFSA, García Fierro, R., Thomas-Lopez, D., Deserio, D., Liebana, E., Rizzi, V., Guerra, B. (2018). Outcome of EC/EFSA questionnaire (2016) on use of whole genome sequencing (WGS) for food- and waterborne pathogens isolated from animals, food, feed and related environmental samples in EU/EFTA countries. *EFSA Supporting Publications*, 15(6), 1432E.

Eijkelkamp, J.M., Aarts, H.J.M., Van Der Fels-Klerx, H.J. (2009). Suitability of rapid detection methods for *Salmonella* in poultry slaughterhouses. *Food Analytical Methods*, 2(1), 1–13.

Emerson, J.B., Adams, R.I., Román, C.M.B., Brooks, B., Coil, D.A., Dahlhausen, K., Ganz, H.H., Hartmann, E.M., Hsu, T., Justice, N.B., Paulino-Lima, I.G., Luongo, J.C., Lymperopoulou, D.S., Gomez-Silvan, C., Rothschild-Mancinelli, B., Balk, M., Huttenhower, C., Nocker, A., Vaishampayan, P., Rothschild, L.J. (2017). Schrödinger's microbes: Tools for distinguishing the living from the dead in microbial ecosystems. *Microbiome*, 5(1), 86.

Ercole, C., Del Gallo, M., Mosiello, L., Baccella, S., Lepidi, A. (2003). *Escherichia coli* detection in vegetable food by a potentiometric biosensor. *Sensors and Actuators B: Chemical*, 91(1), 163–168.

Ercolini, D., Ferrocino, I., Nasi, A., Ndagijimana, M., Vernocchi, P., La Storia, A., Laghi, L., Mauriello, G., Guerzoni, M.E., Villani, F. (2011). Monitoring of microbial metabolites and bacterial diversity in beef stored under different packaging conditions. *Applied and Environmental Microbiology*, 77(20), 7372–7381.

Esteban-Fernández De Ávila, B., Pedrero, M., Campuzano, S., Escamilla-Gómez, V., Pingarrón, J.M. (2012). Sensitive and rapid amperometric magnetoimmunosensor for the determination of *Staphylococcus aureus*. *Analytical and Bioanalytical Chemistry*, 403(4), 917–925.

European Commission (2002). Regulation (EC) No 178/2002 of the European Parliament and of the Council of 28 January 2002 laying down the general principles and requirements of food law, establishing the European Food Safety Authority and laying down procedures in matters of food safety. OJ L 31, 1.2.2002, 1–24.

European Commission (2005). Commission regulation (EC) No 2073/2005 of 15 November 2005 on microbiological criteria for foodstuffs. OJ L 338, 22.12.2005, 1–26.

European Commission (2007). Commission regulation (EC) No 1441/2007 of 5 December 2007 amending regulation (EC) No 2073/2005 on microbiological criteria for foodstuffs. OJ L 322, 7.12.2007, 12–29.

Evenson, M.L., Hinds, M.W., Bernstein, R.S., Bergdoll, M.S. (1988). Estimation of human dose of staphylococcal enterotoxin A from a large outbreak of staphylococcal food poisoning involving chocolate milk. *International Journal of Food Microbiology*, 7(4), 311–316.

Fung, D.Y.C. (2002). Rapid methods and automation in microbiology. *Comprehensive Reviews in Food Science and Food Safety*, 1(1), 3–22.

Gasanov, U., Hughes, D., Hansbro, P.M. (2005). Methods for the isolation and identification of *Listeria* spp. and *Listeria monocytogenes*: A review. *FEMS Microbiology Reviews*, 29(5), 851–875.

Gómez, D., Azón, E., Marco, N., Carramiñana, J.J., Rota, C., Ariño, A., Yangüela, J. (2014). Antimicrobial resistance of *Listeria monocytogenes* and *Listeria innocua* from meat products and meat-processing environment. *Food Microbiology*, 42, 61–65.

Gómez-Govea, M., Solís-Soto, L., Heredia, N., García, S., Moreno, G., Tovar, O., Isunza, G. (2012). Analysis of microbial contamination levels of fruits and vegetables at retail in Monterrey, Mexico. *Journal of Food Agriculture and Environment*, 10, 152–156.

Guy, R.A., Kapoor, A., Holicka, J., Shepherd, D., Horgen, P.A. (2006). A rapid molecular-based assay for direct quantification of viable bacteria in slaughterhouses. *Journal of Food Protection*, 69(6), 1265–1272.

Hindson, B.J., Ness, K.D., Masquelier, D.A., Belgrader, P., Heredia, N.J., Makarewicz, A.J., Bright, I.J., Lucero, M.Y., Hiddessen, A.L., Legler, T.C., Kitano, T.K., Hodel, M.R., Petersen, J.F., Wyatt, P.W., Steenblock, E.R., Shah, P.H., Bousse, L.J., Troup, C.B., Mellen, J.C., Wittmann, D.K., Erndt, N.G., Cauley, T.H., Koehler, R.T., So, A.P., Dube, S., Rose, K.A., Montesclaros, L., Wang, S., Stumbo, D.P., Hodges, S.P., Romine, S., Milanovich, F.P., White, H.E., Regan, J.F., Karlin-Neumann, G.A., Hindson, C.M., Saxonov, S., Colston, B.W. (2011). High-throughput droplet digital PCR system for absolute quantitation of DNA copy number. *Analytical Chemistry*, 83(22), 8604–8610.

Hnasko, R. (2015). *ELISA: Methods and Protocols*. Humana Press, New York.

Hoffmann, S., Maculloch, B., Batz, M. (2015). Economic burden of major foodborne illnesses acquired in the United States. Report, Department of Agriculture, Economic Research Service.

Holland, D. and Mahmoudzadeh, N. (2020). Foodborne disease estimates for the United Kingdom in 2018. Report, Food Standards Agency.

Jackson, B.R., Tarr, C., Strain, E., Jackson, K.A., Conrad, A., Carleton, H., Katz, L.S., Stroika, S., Gould, L.H., Mody, R.K., Silk, B.J., Beal, J., Chen, Y., Timme, R., Doyle, M., Fields, A., Wise, M., Tillman, G., Defibaugh-Chavez, S., Kucerova, Z., Sabol, A., Roache, K., Trees, E., Simmons, M., Wasilenko, J., Kubota, K., Pouseele, H., Klimke, W., Besser, J., Brown, E., Allard, M., Gerner-Smidt, P. (2016). Implementation of nationwide real-time whole-genome sequencing to enhance listeriosis outbreak detection and investigation. *Clinical Infectious Diseases*, 63(3), 380–386.

Jagadeesan, B., Gerner-Smidt, P., Allard, M.W., Leuillet, S., Winkler, A., Xiao, Y., Chaffron, S., Van Der Vossen, J., Tang, S., Katase, M., McClure, P., Kimura, B., Ching Chai, L., Chapman, J., Grant, K. (2019). The use of next generation sequencing for improving food safety: Translation into practice. *Food Microbiology*, 79, 96–115.

Jasson, V., Jacxsens, L., Luning, P., Rajkovic, A., Uyttendaele, M. (2010). Alternative microbial methods: An overview and selection criteria. *Food Microbiology*, 27(6), 710–730.

Kable, M.E., Srisengfa, Y., Laird, M., Zaragoza, J., McLeod, J., Heidenreich, J., Marco, M.L. (2016). The core and seasonal microbiota of raw bovine milk in tanker trucks and the impact of transfer to a milk processing facility. *mBio*, 7(4), e00836–00816.

Käfferstein, F.K. (2003). Food safety as a public health issue for developing countries. In *Food Safety in Food Security and Food Trade. Brief 2 of 17*, Unnevehr, L.J. (ed.). International Food Policy Research Institute, Washington DC.

Keith, M. (1997). Evaluation of an automated enzyme-linked fluorescent immunoassay system for the detection of *Salmonellae* in foods. *Journal of Food Protection*, 60, 682–685.

Kirk, M.D., Pires, S.M., Black, R.E., Caipo, M., Crump, J.A., Devleesschauwer, B., Döpfer, D., Fazil, A., Fischer-Walker, C.L., Hald, T., Hall, A.J., Keddy, K.H., Lake, R.J., Lanata, C.F., Torgerson, P.R., Havelaar, A.H., Angulo, F.J. (2015). World health organization estimates of the global and regional disease burden of 22 foodborne bacterial, protozoal, and viral diseases, 2010: A data synthesis. *PLOS Medicine*, 12(12), e1001921.

Kruse, H. (ed.) (2015). *The Burden of Foodborne Diseases in the European Union.* WHO, Copenhagen.

Kumar, B.K., Raghunath, P., Devegowda, D., Deekshit, V.K., Venugopal, M.N., Karunasagar, I., Karunasagar, I. (2011). Development of monoclonal antibody based sandwich ELISA for the rapid detection of pathogenic *Vibrio parahaemolyticus* in seafood. *International Journal of Food Microbiology*, 145(1), 244–249.

Lan, Y.-B., Wang, S.-Z., Yin, Y.-G., Hoffmann, W.C., Zheng, X.-Z. (2008). Using a surface plasmon resonance biosensor for rapid detection of *Salmonella* Typhimurium in chicken carcass. *Journal of Bionic Engineering*, 5(3), 239–246.

Lartigue, M.F., Héry-Arnaud, G., Haguenoer, E., Domelier, A.S., Schmit, P.O., Van Der Mee-Marquet, N., Lanotte, P., Mereghetti, L., Kostrzewa, M., Quentin, R. (2009). Identification of *Streptococcus agalactiae* isolates from various phylogenetic lineages by matrix-assisted laser desorption ionization-time of flight mass spectrometry. *Journal of Clinical Microbiology*, 47(7), 2284–2287.

Law, J.W.-F., Ab Mutalib, N.-S., Chan, K.-G., Lee, L.-H. (2015). Rapid methods for the detection of foodborne bacterial pathogens: Principles, applications, advantages and limitations. *Frontiers in Microbiology*, 5(770), 19.

Leonard, P., Hearty, S., Brennan, J., Dunne, L., Quinn, J., Chakraborty, T., O'Kennedy, R. (2003). Advances in biosensors for detection of pathogens in food and water. *Enzyme and Microbial Technology*, 32(1), 3–13.

Leonard, P., Hearty, S., Quinn, J., O'Kennedy, R. (2004). A generic approach for the detection of whole *Listeria monocytogenes* cells in contaminated samples using surface plasmon resonance. *Biosens Bioelectron*, 19(10), 1331–1335.

Leonard, S.R., Mammel, M.K., Lacher, D.W., Elkins, C.A. (2015). Application of metagenomic sequencing to food safety: Detection of Shiga toxin-producing *Escherichia coli* on fresh bagged spinach. *Applied and Environmental Microbiology*, 81(23), 8183–8191.

Leonard, S.R., Mammel, M.K., Lacher, D.W., Elkins, C.A. (2016). Strain-level discrimination of Shiga toxin-producing *Escherichia coli* in spinach using metagenomic sequencing. *PloS One*, 11(12), e0167870.

Leong, D., Nicaogáin, K., Luque-Sastre, L., O'Brien, A.D., Hunt, K., Alvarez-Ordóñez, A., Scollard, J., Schmalenberger, A., Fanning, S., O'Byrne, C., Jordan, K. (2017). A 3-year multi-food study of the presence and persistence of *Listeria monocytogenes* in 54 small food businesses in Ireland. *International Journal of Food Microbiology*, 249, 18–26.

Lindstedt, B.A. (2005). Multiple-locus variable number tandem repeats analysis for genetic fingerprinting of pathogenic bacteria. *Electrophoresis*, 26(13), 2567–2582.

Liu, D., Lawrence, M.L., Austin, F.W., Ainsworth, A.J. (2007). A multiplex PCR for species- and virulence-specific determination of *Listeria monocytogenes*. *Journal of Microbiological Methods*, 71(2), 133–140.

Liu, Y., Cao, Y., Wang, T., Dong, Q., Li, J., Niu, C. (2019). Detection of 12 common food-borne bacterial pathogens by TaqMan real-time PCR using a single set of reaction conditions. *Frontiers in Microbiology*, 10(222), 9.

Love, D.C. and Sobsey, M.D. (2007). Simple and rapid F+ coliphage culture, latex agglutination, and typing assay to detect and source track fecal contamination. *Applied and Environmental Microbiology*, 73(13), 4110–4118.

Maciorowski, K.G., Herrera, P., Jones, F.T., Pillai, S.D., Ricke, S.C. (2006). Cultural and immunological detection methods for *Salmonella* spp. in animal feeds – A review. *Veterinary Research Communications*, 30(2), 127–137.

Mahon, C. and Lehman, D. (2018). *Textbook of Diagnostic Microbiology*, 6th edition. Elsevier, Maryland Heights.

Maiden, M.C., Jansen Van Rensburg, M.J., Bray, J.E., Earle, S.G., Ford, S.A., Jolley, K.A., McCarthy, N.D. (2013). MLST revisited: The gene-by-gene approach to bacterial genomics. *Nature Reviews Microbiology*, 11(10), 728–736.

Mandal, P.K., Biswas, A.K., Choi, K., Pal, U.K. (2011). Methods for rapid detection of foodborne pathogens: An overview. *American Journal of Food Technology*, 6, 87–102.

Margot, H., Zwietering, M.H., Joosten, H., O'Mahony, E., Stephan, R. (2015). Evaluation of different buffered peptone water (BPW) based enrichment broths for detection of Gram-negative foodborne pathogens from various food matrices. *International Journal of Food Microbiology*, 214, 109–115.

Matle, I., Mbatha, K.R., Madoroba, E. (2020). A review of *Listeria monocytogenes* from meat and meat products: Epidemiology, virulence factors, antimicrobial resistance and diagnosis. *The Onderstepoort Journal of Veterinary Research*, 87(1), e1–e20.

Mead, P.S., Slutsker, L., Dietz, V., McCaig, L.F., Bresee, J.S., Shapiro, C., Griffin, P.M., Tauxe, R.V. (1999). Food-related illness and death in the United States. *Emerging Infectious Diseases*, 5(5), 607–625.

Muñoz-Berbel, X., Vigués, N., Jenkins, A.T., Mas, J., Muñoz, F.J. (2008). Impedimetric approach for quantifying low bacteria concentrations based on the changes produced in the electrode-solution interface during the pre-attachment stage. *Biosensors and Bioelectronics*, 23(10), 1540–1546.

Nadon, C.A., Trees, E., Ng, L.K., Møller Nielsen, E., Reimer, A., Maxwell, N., Kubota, K.A., Gerner-Smidt, P. (2013). Development and application of MLVA methods as a tool for inter-laboratory surveillance. *Eurosurveillance*, 18(35), 20565.

Nocker, A. and Camper, A.K. (2009). Novel approaches toward preferential detection of viable cells using nucleic acid amplification techniques. *FEMS Microbiology Letters*, 291(2), 137–142.

Nogva, H.K., Drømtorp, S.M., Nissen, H., Rudi, K. (2003). Ethidium monoazide for DNA-based differentiation of viable and dead bacteria by 5′-nuclease PCR. *BioTechniques*, 34(4), 804–813.

Oliver, J.D. (2010). Recent findings on the viable but nonculturable state in pathogenic bacteria. *FEMS Microbiology Reviews*, 34(4), 415–425.

Owen, M., Willis, C., Lamph, D. (2010). Evaluation of the TEMPO® most probable number technique for the enumeration of *Enterobacteriaceae* in food and dairy products. *Journal of Applied Microbiology*, 109(5), 1810–1816.

Pal, S., Ying, W., Alocilja, E.C., Downes, F.P. (2008). Sensitivity and specificity performance of a direct-charge transfer biosensor for detecting *Bacillus cereus* in selected food matrices. *Biosystems Engineering*, 99(4), 461–468.

Pan, Y. and Breidt Jr., F. (2007). Enumeration of viable *Listeria monocytogenes* cells by real-time PCR with propidium monoazide and ethidium monoazide in the presence of dead cells. *Applied and Environmental Microbiology*, 73(24), 8028–8031.

Papić, B., Pate, M., Henigman, U., Zajc, U., Gruntar, I., Biasizzo, M., Ocepek, M., Kušar, D. (2017). New approaches on quantification of *Campylobacter jejuni* in poultry samples: The use of digital PCR and real-time PCR against the ISO standard plate count method. *Frontiers in Microbiology*, 8, 331.

Peruzy, M.F., Proroga, Y.T.R., Capuano, F., Corrado, F., Santonicola, S., De Medici, D., Delibato, E., Murru, N. (2020). Detection and quantification of *Campylobacter* in foods: New analytic approaches to detect and quantify *Campylobacter* spp. in food samples. *Italian Journal of Food Safety*, 9(2), 8591–8591.

Ponniah, J., Robin, T., Margaret, S., Chai, L.C., Radu, S., Farinazleen, M.G., Cheah, Y.K., Nishibuchi, M., Yoshitsugu, N., Kumar, P., Pradeed, M.K. (2012). Assessment of *Listeria monocytogenes* in salad vegetables through kitchen simulation study. *Journal of Tropical Agriculture and Food Science*, 40, 55–62.

Porcellato, D., Narvhus, J., Skeie, S.B. (2016). Detection and quantification of *Bacillus cereus* group in milk by droplet digital PCR. *Journal of Microbiological Methods*, 127, 1–6.

Rogers, K.R. (2000). Principles of affinity-based biosensors. *Applied Biochemistry and Biotechnology – Part B: Molecular Biotechnology*, 14(2), 109–129.

Rudi, K., Moen, B., Drømtorp, S.M., Holck, A.L. (2005). Use of ethidium monoazide and PCR in combination for quantification of viable and dead cells in complex samples. *Applied and Environmental Microbiology*, 71(2), 1018–1024.

Scallan, E., Hoekstra, R.M., Angulo, F.J., Tauxe, R.V., Widdowson, M.A., Roy, S.L., Jones, J.L., Griffin, P.M. (2011). Foodborne illness acquired in the United States – Major pathogens. *Emerging Infectious Diseases*, 17(1), 7–15.

Singh, A., Poshtiban, S., Evoy, S. (2013). Recent advances in bacteriophage based biosensors for food-borne pathogen detection. *Sensors (Basel)*, 13(2), 1763–1786.

Snowdon, J.A., Buzby, J.C., Roberts, T. (2002). Epidemiology, cost, and risk of foodborne disease. In *Foodborne Diseases*, 2nd edition, Cliver, D. and Rieman, H.P. (eds). Elsevier Press, Amsterdam.

Son, J.R., Kim, G., Kothapalli, A., Morgan, M.T., Ess, D. (2007). Detection of *Salmonella* Enteritidis using a miniature optical surface plasmon resonance biosensor. *Journal of Physics: Conference Series*, 61, 1086–1090.

Stiefel, P., Schmidt-Emrich, S., Maniura-Weber, K., Ren, Q. (2015). Critical aspects of using bacterial cell viability assays with the fluorophores SYTO9 and propidium iodide. *BMC Microbiology*, 15(1), 36.

Trevors, J.T. (2012). Can dead bacterial cells be defined and are genes expressed after cell death? *Journal of Microbiological Methods*, 90(1), 25–28.

Uyttendaele, M., Vanwildemeersch, K., Debevere, J. (2003). Evaluation of real-time PCR vs automated ELISA and a conventional culture method using a semi-solid medium for detection of *Salmonella*. *Letters in Applied Microbiology*, 37(5), 386–391.

Välimaa, A.-L., Tilsala-Timisjärvi, A., Virtanen, E. (2015). Rapid detection and identification methods for *Listeria monocytogenes* in the food chain – A review. *Food Control*, 55, 103–114.

Vaz-Velho, M., Duarte, G., Gibbs, P.A. (2000). Evaluation of mini-VIDAS rapid test for detection of *Listeria monocytogenes* from production lines of fresh to cold-smoked fish. *Journal of Microbiological Methods*, 40(2), 147–151.

Velasco-Garcia, M.N. and Mottram, T. (2003). Biosensor technology addressing agricultural problems. *Biosystems Engineering*, 84(1), 1–12.

Velusamy, V., Arshak, K., Korostynska, O., Oliwa, K., Adley, C. (2010). An overview of foodborne pathogen detection. In the perspective of biosensors. *Biotechnology Advances*, 28(2), 232–254.

Vernozy-Rozand, C., Mazuy-Cruchaudet, C., Bavai, C., Richard, Y. (2004). Comparison of three immunological methods for detecting staphylococcal enterotoxins from food. *Letters in Applied Microbiology*, 39(6), 490–494.

Vidic, J. and Auger, S. (2019). Techniques moléculaires de détection de bactéries d'origine alimentaire. *Les techniques de l'ingénieur*, RE281 v1.

Vieira-Pinto, M., Oliveira, M., Bernardo, F., Martins, C. (2007). Rapid detection of *Salmonella* sp. in pork samples using fluorescent *in situ* hybridization: A comparison with VIDAS®-SLM system and ISO 6579 cultural method. *Arquivo Brasileiro de Medicina Veterinária e Zootecnia*, 59, 1388–1393.

Wang, Y., Yang, M., Liu, S., Chen, W., Suo, B. (2015). Detection of viable *Salmonella* in ice cream by TaqMan real-time polymerase chain reaction assay combining propidium monoazide. *Journal of Food and Drug Analysis*, 23(3), 480–485.

Waswa, J., Irudayaraj, J., Debroy, C. (2007). Direct detection of *E. coli* O157:H7 in selected food systems by a surface plasmon resonance biosensor. *LWT – Food Science and Technology*, 40(2), 187–192.

WHO and World Organization for Animal Health (2014). *Listeria monocytogenes*. OIE Terrestrial Manual.

Wu, V.C.H. (2008). A review of microbial injury and recovery methods in food. *Food Microbiology*, 25(6), 735–744.

Yang, X., Noyes, N.R., Doster, E., Martin, J.N., Linke, L.M., Magnuson, R.J., Yang, H., Geornaras, I., Woerner, D.R., Jones, K.L., Ruiz, J., Boucher, C., Morley, P.S., Belk, K.E. (2016). Use of metagenomic shotgun sequencing technology to detect foodborne pathogens within the microbiome of the beef production chain. *Applied and Environmental Microbiology*, 82(8), 2433–2443.

Zhang, G. (2013). Foodborne pathogenic bacteria detection: An evaluation of current and developing methods. *The Meducator*, 1(24), 27–30.

Zhang, W., Jayarao, B.M., Knabel, S.J. (2004). Multi-virulence-locus sequence typing of *Listeria monocytogenes*. *Applied and Environmental Microbiology*, 70(2), 913–920.

Zhao, X., Lin, C.-W., Wang, J., Oh, D.H. (2014). Advances in rapid detection methods for foodborne pathogens. *Journal of Microbiology and Biotechnology*, 24(3), 297–312.

3

Exposure Assessment of Microbial Pathogens

Maarten NAUTA
Statens Serum Institut, Copenhagen, Denmark

3.1. Introduction/state of the art

3.1.1. *Exposure assessment in microbiological risk assessment*

Microbiological risk assessment has predominantly been applied in the context of food and water safety. In quantitative microbiological risk assessment (QMRA), the exposure assessment aims to describe to what extent consumers are ingesting the identified microbiological hazard. The term "exposure" originates from a broader perspective of risk assessment involving chemicals and environmental risks and may suggest that the hazard can enter the body via inhalation or via the skin as well, but this is not considered a relevant route of exposure for foods and water. The health risks considered in the risk assessment are a direct consequence of this exposure. More precisely, the health risks, which are assessed through the dose–response relation used in the hazard characterization, depend on the frequency of the ingestion of the hazard as well as the ingested doses. Thus, to estimate the human health risks, both the frequency of exposure and the ingested doses have to be estimated. This is the primary objective of exposure assessment.

The main challenge of exposure assessment for microbiological risk assessment is a consequence of the fact that ingested doses cannot be

Microbiological Risk Assessment Associated with the Food Processing and Distribution Chain, coordinated by Jeanne-Marie MEMBRÉ.

measured directly. It is simply not feasible to sample and test what people have on their plates when they consume their foods, certainly not in a domestic settings where most food is consumed. In practice, samples can only be taken in the food production and food preparation steps before the food is consumed; that is, during food processing, at retail and in professional kitchens. The problem is then that data obtained at these occasions are not necessarily representative for the ingested doses: after the point where samples are taken, growth can occur during storage of the food, and inactivation and cross-contamination can occur during food preparation. The effect of these processes on the numbers of pathogenic microorganisms in the food should be known. The only way to estimate the ingested doses is therefore to develop and use a mathematical model that describes growth, inactivation and potential transfer of the microorganisms in all processes taking place between the point where data are available and the actual exposure.

There are two more arguments why microbiological exposure assessment requires the modeling of food production and food preparation in a so-called "food pathway" or "exposure pathway". The first is illustrated in Figure 3.1. It is easier to obtain data on the occurrence of the hazard at primary production (the "farm" stage when the food animals are raised or the crops are grown), for which the prevalence and concentrations of the hazard are higher. For example, fewer samples have to be taken to obtain a representative dataset on the occurrence of *Salmonella* in pig herds than for one on its occurrence in pork sausages. However, to apply those data, the occurrence in pigs has to be translated into ingested doses, which may require a complex model describing the transfer, growth and survival over the whole pathway from "farm to fork", with many uncertainties along the way. Data from pork sausages are taken closer to the consumption of the food and would therefore be more representative for the exposure. Further down the pathway, prevalences and concentrations of the hazard usually decrease. More samples have to be taken to get good estimates of prevalence and concentrations; the pathway to the actual exposure is shorter and may be easier to describe, with less uncertainties. It is therefore important to find the right balance between data availability and model complexity when deciding on the starting point of the food pathway used for the exposure assessment.

A second argument for modeling a long food pathway lies in the objective of the risk assessment. Often, it is not only important to estimate the current public health risk, but also to evaluate and compare the expected impact of control measures that are taken along the food pathway. If such control measures are to be taken at the farm, obviously, the farm stage has to be included in the food pathway. If such control measures are not considered, a shorter pathway may be used.

Overall, the objectives of the risk assessment and the availability of data are the basis for choosing the methodology to be used in exposure assessment. After considering these, drafting the outline of the food pathway, including the choice of the starting point and the routes of transmission of the hazard, is the first step of the process (Nauta 2008).

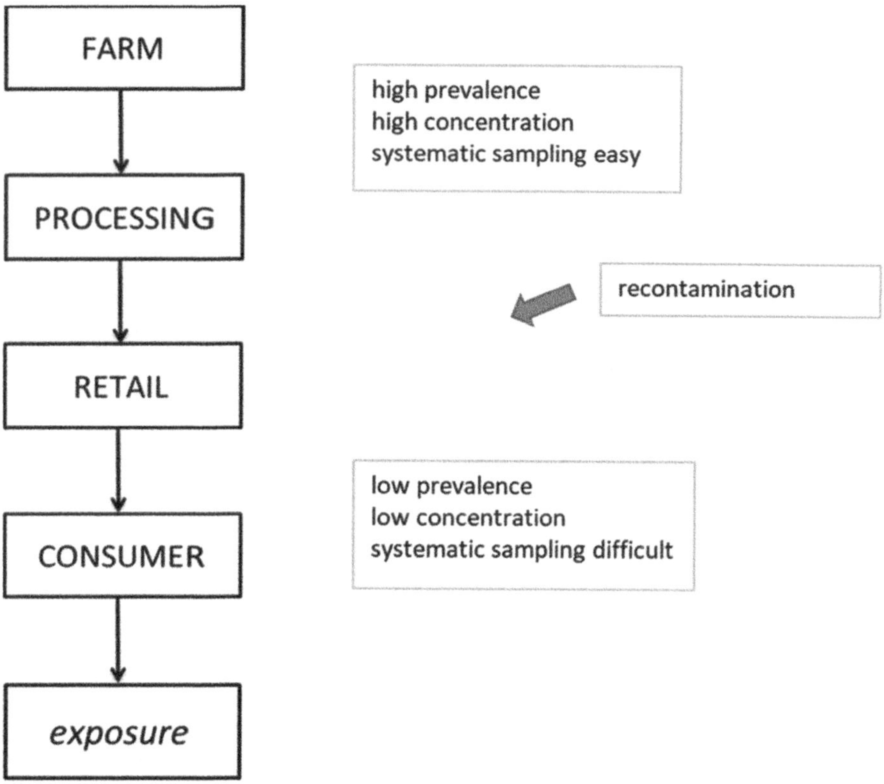

Figure 3.1. *In the farm-to-fork food pathway, prevalence and concentrations are usually higher closer to the farm, so it is easier to obtain data there. However, a model covering a long pathway is needed to estimate exposure, and recontamination during the process may complicate the modeling*

3.1.2. *The output of an exposure assessment*

Exposure assessment is usually performed as part of a QMRA study, where the mean risk for a predefined population of consumers is assessed. Within a QMRA, the exposure assessment is linked into the dose–response relation by providing the doses ingested by the population of consumers, in terms of (1) the frequency of exposure and (2) the ingested doses themselves. As microbiological risks are usually about acute effects following a single exposure, risk estimates are based on risks per serving: the consumption of a certain amount of the food product of concern may result in an effect, that is, infection and/or illness.

To calculate the risk per serving, we also need to have an estimate of the dose per serving. One serving with one dose is associated with a single estimate of the probability of illness (the risk). For all servings consumed by the population of consumers, there is a set of different doses to which the consumers are exposed, usually expressed as a distribution providing the variability of ingested doses. In principle, a point estimate of the ingested dose, such as the mean dose, along with an estimate of the likelihood of that dose (the frequency of exposure), is useful to estimate the risk by means of a dose–response model. However, this risk estimate need not be the same as the mean risk estimate that is obtained when the variability between ingested doses is included. The reason is that the probability of illness is often not linearly related with the dose, so that high doses have a different weight than low doses. Only if the dose–response relation can be assumed to be linear will knowledge of the arithmetic mean of the dose be sufficient to characterize the exposure.

Therefore, for each serving, we need an estimate of the probability that it is contaminated with the hazard (i.e. that the dose is larger than zero). This probability of contamination is often referred to as the true prevalence of the hazard in contaminated products. If a serving is contaminated, we need an estimate of the ingested dose. The ingested dose varies between servings due to the variation in the initial concentrations at the start of the food pathway and the variation in process parameters, such as processing times and the temperatures applied during storage and cooking. In general, it is important to describe this variability in doses as a result of the exposure assessment, as this variability impacts the estimate of the mean population risk.

3.1.3. *Approaches for analyzing food-pathway models*

The first food-pathway model developed for quantitative microbiological risk assessment was published by Cassin *et al.* (1998) for *Escherichia coli* O157:H7 in ground beef hamburgers. The authors introduced the process risk model (PRM) to "describe the integration and application of the quantitative risk assessment methodology with scenario analysis and predictive microbiology [and thus] provide an objective assessment of the hygienic characteristics of a manufacturing process". They constructed a stochastic model where changes in prevalence and concentration were followed, starting with the fecal contamination of the beef carcass. A Monte-Carlo simulation was performed to reflect hypothetical scenarios in terms of the values attributed to the identified factors in the exposure assessment. The model published by Cassin *et al.* (1998) has long served as a blueprint for QMRA.

The PRM methodology was later extended and structured by Nauta (2001, 2002, 2008) through the development of the modular process risk model (MPRM) approach. In the MPRM approach, the specific process step in the PRM is characterized by the impact on the food product and the microorganisms. It describes not only the change in prevalence and concentration, but also the change in "unit size", that is, the weight or volume of the food product as it is defined in the production process. This led to the identification of six "basic processes", two microbial processes (growth and inactivation) and four food-handling processes (mixing, partitioning, cross-contamination and removal).

The microbial processes depend on the microbial agent (the hazard) of concern and can be described by predictive microbiology models that are specific for the hazard. The other processes depend on the way the food is handled and impacted by the specific hazard concerned. Mixing and partitioning describe how food units are joined (e.g. beef trimmings mixed into a large batch of ground beef or milk from different cows combined into a milk tank) or split up (e.g. a large batch of beef split up in smaller packages or milk from a tank distributed across several milk packages) (Nauta 2005). Cross-contamination involves a combination of transfer events that leads to contamination of foods, sometimes even in the case of a single bacterial transfer event (Pérez-Rodríguez *et al.* 2008; Møller *et al.* 2021). Removal refers to the selective removal of food items based on a (visual) inspection of their quality, which may be related to the probability and level of contamination of this food item.

An advantage of the MPRM approach is that a limited number of models need to be developed and they can be re-used at different steps in the food pathway, which increases the transparency of the model. In addition, it allows for a first overview of the impact that the different steps in the food pathway have on the changes in prevalence, concentration and unit size, as exemplified in Table 3.1. A good example of the application of the MPRM approach can be found in Nauta *et al.'s* (2001) risk assessment of Shiga toxin-producing *Escherichia coli* O157 in steak tartare in the Netherlands, highlighting a food pathway from cattle to steak tartare patties (see Figure 3.2).

The MPRM approach has been applied in FDA-iRisk[1], a web-based system designed to analyze data concerning microbial and chemical hazards in food that allows the user to construct food pathway models and perform QMRA. In FDA-iRisk, some additional basic processes are defined and a somewhat different nomenclature is applied. It is a highly suitable tool for developing, analyzing and storing food-pathway models.

	Effect on P	Effect on N	Effect on U
Growth	=	+	=
Inactivation	-	-	=
Mixing	+	=	+
Partitioning	-	=	-
Removal	-	-	=
Cross-contamination	+	-/=/+	=

Table 3.1. *Basic processes and their effect on the prevalence (P; fraction of contaminated units), number of organisms in all units (N) and the unit size (U). =: no effect; +: increase; -: reduction. Adapted from Nauta (2002)*

Food pathway models constructed in the development of a PRM or MPRM may be complex and data-demanding. Often, these models use stochastic modeling, incorporating the variability and/or uncertainty associated with the events modeled. If properly applied, this increases the accuracy and transparency of the risk assessment, but it is not always necessary or feasible to do it. A simplified approach, such as the one defined

1 Available at: https://irisk.foodrisk.org/ (accessed February 2021)

by the International Commission on Microbiological Specifications for Foods (ICMSF) (Zwietering *et al.* 2010), may be sufficient as well. This approach has primarily been developed to check whether a food production process complies with a Food Safety Objective (FSO), but it can be applied for a quick exposure assessment as well.

The ICMSF equation builds on concentrations expressed as log CFUs/g, so the exposure can be expressed as the sum of the effects of growth, inactivation and contamination:

$$H_0 + \sum G + \sum C - \sum R = exposure$$

where H_0 is the initial concentration in a food product, ΣG is the sum of the effect of growth in the process steps that allow growth, ΣR is the sum of the effect of inactivation (reduction) in the process steps that lead to inactivation and ΣC is the increase due to (re-)contamination.

This equation is useful to check whether the concentrations that can be found after a food manufacturing process are likely to be acceptable or not (Zwietering and Nauta 2007). It requires predictive models that describe growth and inactivation, for example, through the use of a software tool like Combase (www.combase.cc) or FSSP (fssp.food.dtu.dk) (see also Tenenhaus-Aziza and Ellouze (2015)), but it does not require specific knowledge on QMRA. With a relatively quick analysis, it can help a food manufacturer to evaluate whether a production process is safe or to decide that a more advanced risk assessment is needed. Disadvantages of the use of this equation include the following:

– it cannot deal with mixing, partitioning and removal;

– re-contamination is not additive on a log scale, like growth and inactivation (Møller *et al.* 2021);

– variability is not included, neither in the inputs nor the output.

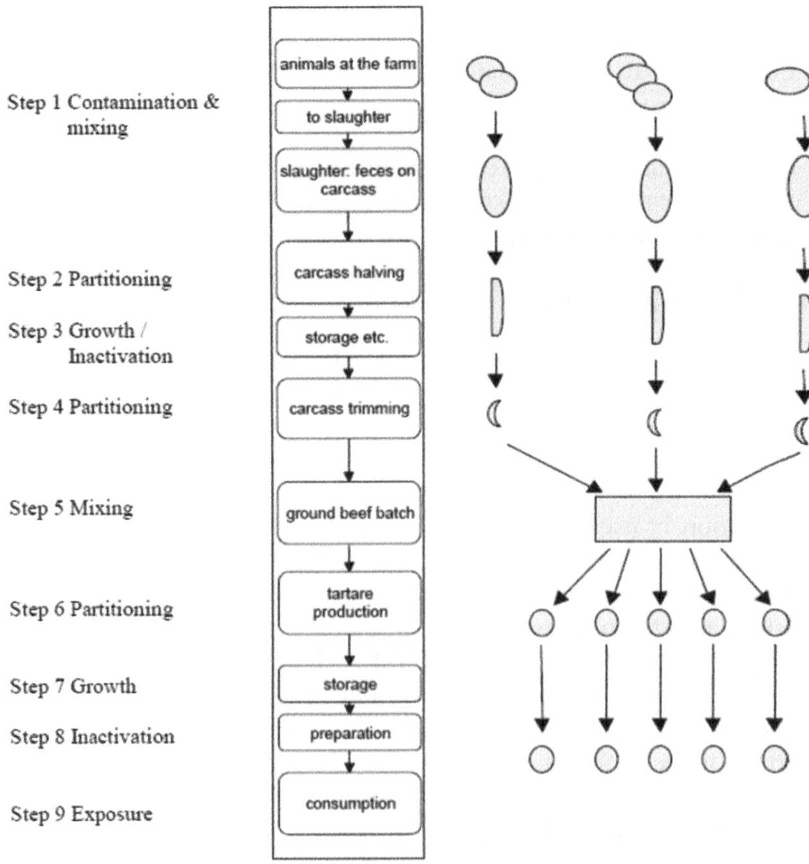

Figure 3.2. *A schematic representation of the food pathway for a QMRA of* E. coli *O157 in beef tartare (Nauta* et al. *2001). From top down, several slaughter animals (lying ovals) contaminate carcasses (hanging ovals), which after halving and trimming contaminate a ground beef batch (rectangle). From the ground beef batch, tartare patties are produced (circles). Basic process steps 1–9 are identified as the basis for a series of consecutive modules*

3.2. Construction of a food-pathway model

This section describes an example of a food-pathway model for exposure assessment in order to illustrate how such a model can be built. In the example, we consider the exposure to *Salmonella* spp. through the consumption of meatballs made of pork, similar to the *frikadeller* that are popular in Denmark. These meatballs are made of ground pork, sold at retail and fried by the consumer. They may be directly eaten after frying, but in the example they are stored afterward and then consumed without further heating, which is a common habit for many consumers.

The exposure assessment is performed in three steps, as explained below:

– the food pathway is drawn, the processes involved are describe and the models to use for each process are defined;

– a deterministic food-pathway model is analyzed;

– a stochastic food-pathway model is analyzed.

The purpose of the example is to illustrate which activities are involved in exposure assessment and how these can result in the actual assessment. Building up a food-pathway model for exposure assessment following these three steps is always recommended. Several simplifying assumptions are made for the sake of the example. It is not meant to be a realistic example of an exposure assessment of *Salmonella* in meatballs to be used for risk management in practice.

3.2.1. *Description of the food pathway*

The starting point of the food pathway is pork cuts that are produced at a slaughterhouse. Some of these pork cuts are contaminated with *Salmonella.* In a grinder, these pork cuts are ground to produce a batter of minced pork, which is used to make a large number of meatballs. These meatballs are sold at retail and transported, stored and prepared by the consumer. After preparation, they may be eaten directly or stored and eaten cold later. The questions are (1) what percentage of meatballs are contaminated with *Salmonella* at the moment of consumption (i.e. the frequency of exposure) and (2) what is the distribution of the ingested doses from the contaminated meatballs?

An outline of the food pathway is given in Figure 3.3. It shows the different steps in the process that may be relevant to consider in the food pathway model.

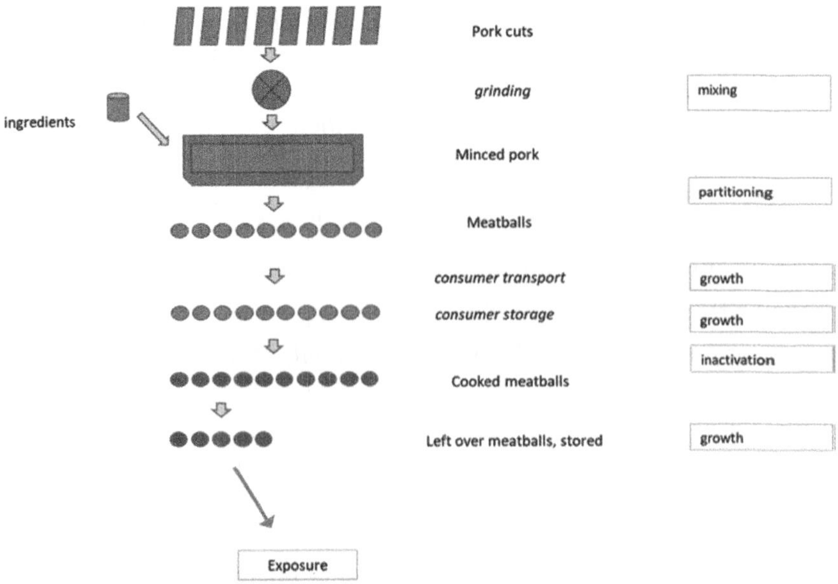

Figure 3.3. *Food pathway for Danish meatballs. Meatballs are produced from minced pork; the basic processes of the process steps are identified*

Once the steps in the food pathway are defined, basic processes are assigned to the steps. These are also shown in Figure 3.3. First, the grinding of pork cuts is a mixing process, where pork cuts are combined into a large batter of minced pork. The minced pork batter is then divided into meatballs, which is a partitioning process. Meatballs are stored at retail and sold, presumably under controlled conditions that prevent growth. Consumers buy the meatballs; their storage during transport and at home is no longer under regulatory control, which implies that growth may occur. Frying will inactivate the *Salmonella*, but some may survive. If the meatballs are stored under uncontrolled conditions again, growth may occur. The latter is the scenario considered in the exposure assessment.

As shown in Figure 3.3, this food pathway model includes basic processes mixing, partitioning, growth and inactivation.

3.2.2. *Step 2: deterministic model*

For the deterministic model, predictive models can be selected to model bacterial growth and inactivation, while other models should be defined for processes such as mixing, partitioning, cross-contamination and removal. Typically, a deterministic model uses point estimates for the parameter values and does not include stochasticity, variability or uncertainty.

3.2.2.1. *Mixing and partitioning models*

In the example, the first two process steps are the mixing of pork cuts into a batch of minced meat and the partitioning of the minced meat in the batch into meatballs. The mixing results in an increase in unit size and consequently an increase in prevalence and a decrease in the concentration in the contaminated units. As there is no growth or inactivation, the total number of bacteria in all the meat (contaminated and uncontaminated) remains the same. Similarly, due to stochasticity, the partitioning may result in a decrease in prevalence and increase in concentration, whereas the total number of bacteria remains the same. The models applied are taken from Nauta (2005 and 2008).

For the grinding, the prevalence of contaminated batches (P_{bat}) depends on the prevalence of contaminated pork cuts (P_{cut}), the weight of the pork cuts (W_{cut} = 1 kg) and the weight of a minced meat batch (W_{bat} = 50 kg):

$$P_{bat} = 1 - (1 - P_{cut})^{Wbat/Wcut}$$

The concentration (in log CFUs/g) in contaminated batches (C_{bat}) depends on the concentration in the contaminated pork cuts (C_{cut}) and the prevalences:

$$C_{bat} = C_{cut} + log_{10}(^{P_{cut}}/_{P_{bat}})$$

After making the meatballs, the prevalence in the meatballs, P_{mba}, is

$$P_{mba} = P_{bat}(1 - \exp(-1000\ W_{mba} 10^{C_{bat}}))$$

The weight of a single meatball is assumed to be W_{mba} = 50 g.

Then, the concentration in the meatballs (C_{mba}) becomes

$$C_{mba} = C_{bat} + log_{10}(^{P_{bat}}/_{P_{mba}})$$

Note that these models assume that the meat is well mixed and that all contaminated meatballs are contaminated with the same concentrations. This is clearly a simplification of the stochastic processes of mixing and partitioning.

3.2.2.2. *Predictive models*

The selection of the most suitable predictive models depends on the hazard and food product of concern, as well as the specific demands of the exposure assessment. In general, it is best to select models that are specific for the hazard and the food product (in this case, *Salmonella* in pork). If these are not available, we have to select appropriate surrogates. Such a surrogate may be a similar food product (such as chicken meat instead of pork), but it is also possible to base the models on more generic characteristics of the food. Here, it is usually relevant to know the salt content (water activity, a_w), the acidity (pH) and possibly other characteristics that influence the growth, which are often included as parameters in secondary growth models. Although it may be attractive to apply predictive modeling software (Tenenhaus-Aziza and Ellouze 2015), using the selected predictive model in the exposure assessment often demands that the model is implemented by the risk assessor. This facilitates the use in the context of the exposure assessment. If several models are available for the hazard and food product considered, the risk assessor should clarify its choice in the model description.

In the example, for the growth model, we apply the model used by Møller *et al.* (2015) in their risk assessment of *Salmonella* in Danish meatballs produced in the catering sector. This model was selected as it has been published in the peer-reviewed scientific literature for a risk assessment of the same hazard in the same food product. In addition, it contains the relevant parameters (time and temperature) and the equation is well-described in their paper, which allows re-implementation in an exposure assessment model.

The growth model used is an exponential growth model:

$$\log(N_t) = \log(N_0) + \frac{\mu_{max}\,(t - lag)}{\ln(10)}$$

where N_t is the concentration after *t* hours, N_0 is the initial concentration and *lag* is the lag time. If $t < lag$, $\log(N_t)=\log(N_0)$.

The maximum growth rate is defined as:

$$\sqrt{\mu_{max}} = b(T - T_{min})\,(1 - \exp(c(T - T_{max})))$$

with constants b =0.04 $h^{-0.5}{}^{\circ}C^{-1}$ and c = 0.43°C^{-1}. T is the temperature, and T_{min} (3.94°C) and T_{max} (48.0°C) are the minimum and maximum theoretical temperatures for growth, respectively.

The lag time (h) is applied for each of the different steps where the growth model is applied. As in Møller *et al.* (2015), it is defined as:

$$lag = \frac{RLT \times \ln(2)}{\mu_{max}}$$

with relative lag time RLT = 3.1.

The growth model uses the concentration at the end of the previous step as the initial concentration, and is characterized by time t and temperature T. It is applied for transport to home (*trans*), domestic storage before cooking (*stor1*) and storage after cooking (*stor2*).

The inactivation model used is Bigelow's model:

$$\log(N_t) = \log(N_0) - \frac{t}{D_T}$$

where

$$D_T = D_{ref} 10^{\frac{Tref - T}{z}}$$

and where D_T is the decimal reduction time at temperature T, D_{ref} is the reference decimal reduction time at temperature T_{ref} (D_{ref} = 5.2 min at T_{ref} = 60°C) and z is the z value (z = 7.34°C).

The inactivation model is used for the effect of cooking (*cook*) between the two storage steps, characterized by the time t and temperature T.

3.2.2.3. *Model implementation*

The model can be easily implemented in Excel and in programming software. As an example, the R code is presented in Box 3.1. The model output is the ingested dose D (in CFUs), which is calculated as the concentration after second storage multiplied by the weight of a single meatball (W_{mba}). An underlying assumption is that only one meatball in a portion is contaminated, which makes sense if the prevalence of contaminated meatballs is low and contaminated meatballs are randomly distributed.

The prevalence of potentially contaminated portions with portion size *ps* is calculated as:

$$P_{exp} = 1 - (1 - P_{mba})^{ps/Wmba}$$

A deterministic model typically presents one of many possible "what if" scenarios: it provides the exposure that would occur given the values applied for the model parameters. In this case, it is assumed that the production conditions are well-controlled by the food business operator and are well-known. The conditions applied by the consumer and the initial concentration of the product may vary. Therefore, the modeled exposure is representative for a scenario that is defined by the given initial prevalence and concentration; time and temperature conditions during transport, storage and cooking; and portion size. The most realistic scenario is the "baseline" scenario, for which the model parameters have values that are considered "most likely". Values for the baseline are given in Table 3.2.

When implemented in the model, the baseline scenario gives a prevalence of 18.1% contaminated portions, but the dose in all those portions is 0 CFUs. Therefore, even all the portions derived from contaminated meat (18.1%) are uncontaminated. The reason for this is that all *Salmonella* are inactivated with a cooking step taking five minutes at 70°C. This implies that, if the meatballs are properly stored and cooked, no consumer is exposed to *Salmonella*.

	Parameter	Value	Unit
Prevalence in pork cuts	P_{cut}	0.2	%
Concentration in pork cuts	C_{cut}	2	Log CFUs/g
Transport temperature	T_{trans}	10	°C
Transport time	t_{trans}	1	h
Storage temperature	T_{stor1}	6	°C
Storage time	t_{stor1}	24	h
Cooking temperature	T_{cook}	70	°C
Cooking time	t_{cook}	5	min
Storage temperature	T_{stor2}	6	°C
Storage time	t_{stor2}	24	h
Portion size	ps	100	g

Table 3.2. *Baseline model parameters*

This example shows that the most likely scenario is not so interesting for QMRA: food safety risks are usually a consequence of mistakes that result in deviations from good manufacturing practices (GMPs) or good hygiene practices (GHPs). It also shows that it is relevant to explore other scenarios. It is of interest to study the consequences of high initial contamination levels and/or incorrect storage and cooking or other processing conditions.

Table 3.3 illustrates how a deterministic model can be used for scenario analysis, exploring the impact of potential changes in the model parameter values. It shows that the exposure is especially sensitive for the cooking temperatures: if the cooking temperature is high enough, there is no option for the pathogen to survive and grow. With cooking temperatures of 63°C or lower, exposure occurs, and storage conditions become important and can influence the exposure. Initial concentration and prevalence are also important; portion size affects both the probability of exposure and the exposure itself. The program code in Box 3.1 can be used to explore additional alternative scenarios.

Scenario	Deviation from baseline	Exposure (CFUs) to contaminated portions	P_{exp} (%)
Baseline (see Table 3.2)	None	0	18.1
Bad storage conditions	$T_{stor2} = 10$; $t_{stor2} = 48$	0	18.1
Bad cooking conditions	$T_{cook} = 63$; $t_{cook} = 4$	1.1	18.1
Very bad cooking conditions	$T_{cook} = 60$; $t_{cook} = 3$	27.8	18.1
Bad storage and bad cooking conditions	$T_{stor2} = 10$; $t_{stor2} = 48$; $T_{cook} = 63$; $t_{cook} = 4$	1.3	18.1
Bad transport and bad cooking conditions	$T_{trans} = 20$; $t_{strans} = 8$; $T_{cook} = 63$; $t_{cook} = 4$	1.9	18.1
High initial concentration and bad cooking conditions	$C_{cut}=3$; $T_{cook} = 63$; $t_{cook} = 4$	11.2	18.1
High initial prevalence and bad cooking conditions	$P_{cut} = 0.5\%$; $T_{cook} = 63$; $t_{cook} = 4$	1.2	39.4
Large portion size and bad cooking conditions	$ps = 200$; $T_{cook} = 63$; $t_{cook} = 4$	1.1	33.0

Table 3.3. *Model scenarios*

The example shows that a deterministic model is useful to define the models used in the different process steps and to explore the impact of the model parameters on the exposure. This quickly provides the most relevant control points and the steps in the model where good parameter estimates are most useful. However, the model is not particularly useful for a precise estimate of the doses that consumers are exposed to. This requires input on the likelihood of unfavorable conditions, as in a stochastic model.

```
#initial prevalence and concentration
Pcut        <- 0.002
Ccut        <- 2                # log CFUs/g

#weights of cuts, batch and meatballs
Wcut        <- 1                # kg
Wbat        <- 50               # kg
Wmba        <- 0.05             # kg

#serving size
ps  <-100                       # g

#process temperatures (T) and times (t)
#transport
Ttr<-10             # C
ttr<-1              # h
#storage
Tst<-6              # C
tst<-24             # h
#cooking
Tco<-63             # C
tco<-5              # min
#storage2
Tst2<-6             # C
tst2<-24            # h

#growth model for Salmonella (Møller et al.)
growthsalm <- function(t,T,C0)
{
  #growth model parameters
  b<- 0.036
  c<- 0.43
  Tmax<- 48         # C
  Tmin<- 3.9        # C
  #model
  mu        <-(b*(T-Tmin)*(1-exp(c*(T-Tmax))))^2
```

```
 lag        <-3.1*log(2)/mu
 C<-        C0 + mu *(max(0,t-lag))/log(10)
 C
}

#inactivaton model for Salmonella
inactsalm <- function(t,T,C0)
{
 #inactivation model parameters
 Dref<-5.2              # min
 Tref<-60               # C
 z<-7.34                 # C
 #model
 Dval       <- Dref*10^((Tref-T)/z)
 C <- C0-t/Dval
 C
}

#mixing: making a batch of ground meat from pork cuts

Pbat        <-1-(1-Pcut)^(Wbat/Wcut)
Cbat        <-Ccut + log10(Pcut/Pbat)          #log CFUs/g
Pbat
Cbat

#partitioning: making meatballs

Pmba        <-Pbat*(1-exp(-(Wmba*1000*10^Cbat)))
Cmba        <-Cbat + log10(Pbat/Pmba)
Cmba
Pmba

#growth during transport

Ctrans<-growthsalm(ttr,Ttr,Cmba)
Ctrans

#growth during storage

Cstor<-growthsalm(tst,Tst,Ctrans)
Cstor

#inactivation during cooking
```

```
Ccook <- inactsalm(tco,Tco,Cstor)
Ccook

#growth during storage
Cstor2<-growthsalm(tst2,Tst2,Ccook)
Cstor2

#exposure

Pexp <- 1-(1-Pmba)^(ps/(Wmba*1000))
D <-Wmba*1000*10^Cstor2    #assume only one of the meatballs is
contaminated
Pexp
D
```

Box 3.1. *R code of the deterministic model*

3.2.3. *Step 3: stochastic model*

A stochastic model includes variability and/or uncertainty in the model parameters, which are relevant for risk assessment. Variability represents the true heterogeneity of the population of subjects; it is irreducible to additional measurements. In exposure assessment, the actual exposure in a population is always variable, as doses vary between servings, both between individuals in the population and eating occasions. This variability originates from many sources, such as the variability in contamination of the raw materials, storage conditions, and food handling practices and portion sizes, as well as random stochastic variation. Uncertainty represents a lack of perfect knowledge and can be reduced by gathering additional knowledge. Uncertainty can be expressed by subjective probabilities (EFSA Scientific Committee *et al.* 2018). In exposure assessment, the uncertainty is generally large, as there are many unknowns related to the production process, especially in the consumer stage. Also, data and models are imperfect, resulting in imperfect model estimates.

Ideally, both variability and uncertainty are included in a stochastic exposure assessment model. This may require the development of a second-order Monte Carlo simulation model, in which uncertainty and variability are separated. With the separation of variability and uncertainty, it is possible to show how the risk estimate is affected by the variability, and how the uncertainty of the risk estimate can be affected by the uncertainty. More details on this can be found elsewhere (Nauta 2000, 2007; Pouillot and Guillier 2021).

In this chapter, only variability is considered in the stochastic model of *Salmonella* in meatballs made of pork. Table 3.4 shows how the variability is implemented in the model inputs. In this illustrative example, the selected probability distributions are based on assumptions; for a realistic example, they should be based on data collected at the different steps of the process. As the distributions represent variability, they all imply that the model parameters vary randomly and independently between the iterations of the model, as indicated by the distributions. Every iteration represents a scenario where one serving is ingested. It is assumed that we know the distributions, and that there is no uncertainty about them. In addition, it is assumed that the distributions are not correlated. Note that the prevalence is not variable. This should usually be the case in all exposure assessments, as the prevalence is a metric for the population (0.2% of all pork cuts are contaminated with *Salmonella*), whereas concentrations, times and temperatures vary between the individual units and servings. The R code for the stochastic model is given in Box 3.2.

	Parameter	Distribution	Unit
Prevalence in pork cuts	P_{cut}	0.2	%
Concentration in pork cuts	C_{cut}	Normal (2,1)	Log CFUs/g
Transport temperature	T_{trans}	Normal (10,2)	°C
Transport time	t_{trans}	Gamma (1,1)	h
Storage temperature	T_{stor1}	Normal (6,1)	°C
Storage time	t_{stor1}	Gamma (24,1)	h
Cooking temperature	T_{cook}	Normal (70,4)	°C
Cooking time	t_{cook}	Normal (5,1)	min
Storage temperature	T_{stor2}	T_{stor}^{*}	°C
Storage time	t_{stor2}	Gamma (24,1)	h
Portion size	ps	Normal (100,10)	g

*Assuming the same temperature as the first storage.

Table 3.4. *Baseline model parameters*

If variability is added to the model, the resulting model can be used with Monte-Carlo simulations, which basically perform an iterative analysis of a set of different scenarios that present possible exposures. It is important to formulate the model in such a way that each model iteration represents a realistic scenario, that is, a single exposure as it may take place in reality, calculated through random samples of the variable parameters.

The result of the model is a distribution of doses (Figure 3.4). Usually, in a realistic scenario, the vast majority of exposures are close to a dose of zero (fortunately, people are rarely exposed to pathogens; otherwise, that food product would not be on the market). As the distribution of doses is often skewed, this distribution is frequently shown on a log scale. Note that if the logarithm of a dose is zero, this implies exposure to 1 CFU. A dose of 0 CFUs cannot be expressed in logs.

Apart from a graph of the distribution, the results of the exposure assessment can be expressed through several statistics, as illustrated by the example. The mean dose, including zeroes, is 1.75 CFUs; the (10 based) log of the mean dose is then log(1.75) = 0.24 log CFUs. The prevalence of exposure is the percentage of contaminated servings: P_{exp} = 1.2%. Note that this is an increase compared to the prevalence of contaminated pork cuts due to the mixing process during grinding and the consumption of more than one meatball. The mean of the 1.2% positive doses is 141.5 CFUs, whereas the log of the mean of the positive doses is log(141.5) = 2.15 log CFUs, and the mean log dose (of positive doses) is 0.89 log CFUs.

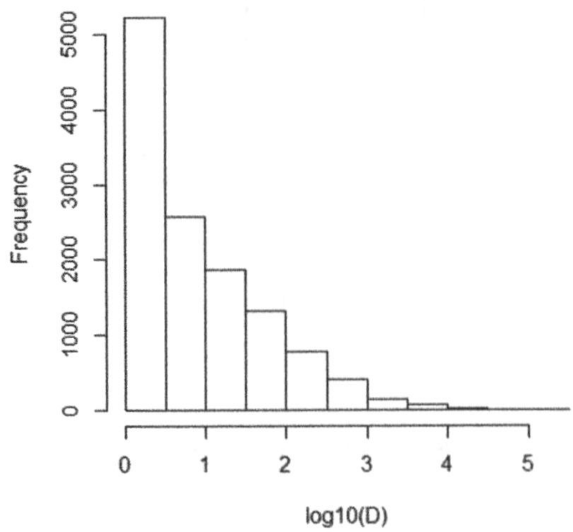

Figure 3.4. *The result of the exposure assessment, showing the frequencies of the doses expressed as log (D) among 1,000,000 model iterations, equivalent to 1,000,000 servings. Note that only contaminated servings appear in the plot, as the dose is expressed in log units*

As with the deterministic model, several stochastic model scenarios can be run and compared. In these scenarios, both fixed model parameters and the probability distribution parameters can be modified. A comparison of these scenarios may show which parameters have most impact on the exposure, which may be important for both the uncertainty of the exposure assessment and the potential impact of process interventions that aim to reduce the exposure.

Through the development of a stochastic model that includes the variability, we obtain a more realistic exposure assessment. It shows that some unlikely scenarios resulting in very high doses may occur due to rare combinations of, for example, high initial concentrations, bad storage conditions, improper inactivation and large portion sizes. In practice, such rare combinations may be those where the consumer gets ill and therefore they are of crucial importance for the risk assessment. For that reason, stochastic exposure assessment models are to be preferred over deterministic ones.

```
sims <- 1000000
#initial prevalence and concentration
Pcut        <- 0.002
Ccut        <- rnorm(sims,2,1)                # log CFUs/g

#weights of cuts, batch and meatballs
Wcut        <- 1              # kg
Wbat        <- 50             # kg
Wmba        <- 0.05           # kg

#serving size
ps  <-rnorm(sims,100,10)                          # g

#process temperatures (T) and times (t)
#transport
Ttr<-rnorm(sims,10,2)                # C
ttr<-rgamma(sims,1,1)                # h
#storage
Tst<-rnorm(sims,6,1)          # C
tst<-rgamma(sims,24,1)               # h
#cooking
Tco<-rnorm(sims,70,4)                # C
tco<-rgamma(sims,5,1)                # min
#storage2
Tst2<-Tst                     # C
tst2<-rgamma(sims,24,1)       # h
```

```
#growth model for Salmonella (Møller et al.)
growthsalm <- function(t,T,C0)
{
 #growth model parameters
 b<- 0.036
 c<- 0.43
 Tmax<- 48          # C
 Tmin<- 3.9         # C
 #model
 mu        <-(b*(T-Tmin)*(1-exp(c*(T-Tmax))))^2
 lag       <-3.1*log(2)/mu
 C<-       C0 + mu *(max(0,t-lag))/log(10)
 C
}

#inactivaton model for Salmonella
inactsalm <- function(t,T,C0)
{
 #inactivation model parameters
 Dref<-5.2          # min
 Tref<-60           # C
 z<-7.34             # C
 #model
 Dval      <- Dref*10^((Tref-T)/z)
 C <- C0-t/Dval
 C
}

#mixing: making a batch of ground meat from pork cuts
Pbat       <-1-(1-Pcut)^(Wbat/Wcut)
Cbat       <-Ccut + log10(Pcut/Pbat)         #log CFUs/g
#partitioning: making meatballs
Pmba       <-Pbat*(1-exp(-(Wmba*1000*10^Cbat)))
Cmba       <-Cbat + log10(Pbat/Pmba)
mean(Cmba)

#growth during transport
Ctrans<-growthsalm(ttr,Ttr,Cmba)
mean(Ctrans)

#growth during storage
Cstor<-growthsalm(tst,Tst,Ctrans)
mean(Cstor)
```

```
#inactivation during cooking
Ccook <- inactsalm(tco,Tco,Cstor)
mean(Ccook)

#growth during storage
Cstor2<-growthsalm(tst2,Tst2,Ccook)
mean(Cstor2)

#exposure
#number of contaminated balls per portion
balls <- ceiling(ps/(Wmba*1000))
posballs <- balls
for (i in 1:sims) posballs[[i]] <-rbinom(1,balls[[i]],Pmba[[i]])

#all ingested doses
#assume only one of the meatballs is contaminated
D  <-Wmba*posballs*1000*10^Cstor2
D <- rpois(sims,D)

posD <- D[ D != 0 ]
#mean dose
mean(D)
#prevalence of exposure
1-length(which(D==0))/sims
#mean log dose (of positives)
mean(log10(posD))
# log mean dose
log10(mean(D))

hist(log10(D))
```

Box 3.2. *R code of the stochastic model*

3.3. Conclusion

3.3.1. *The use of exposure assessment*

Exposure assessment is predominantly applied with the QMRA framework, where it is integrated with a dose–response relation to perform a risk assessment. As microbiological growth and inactivation, together with food-handling processes, have a large influence on the exposure and a crucial impact on the effect of food hygiene and control measures, the exposure assessment contains the instruments for risk managers to control the risk. Dose–response models and risk characterization are described in other chapters of this book. The reader can also refer to other work for descriptions of the role of exposure assessment in QMRA and recent developments (Haas *et al.* 2014; Membré and Valdramidis 2016; Valdramidis *et al.* 2017; FAO and WHO 2020; Pérez-Rodríguez 2021).

Exposure assessment can also be regarded as a standalone process. Although exposure alone cannot inform on the actual health impact of the exposure on the population, it is clear that a higher exposure (i.e. a higher frequency of exposure and being exposed to a larger number of pathogens) will lead to a higher risk. Moreover, it is also clear that, given the same dose, the frequency of exposure is proportional to the population risk. Therefore, exposure assessment alone can be useful, which is particularly relevant in the absence of a dose–response relation.

There are some examples of microbiological hazards where a dose–response relation for the ingested dose and probability of illness per serving are lacking, such as *Bacillus cereus* and *Yersinia enterocolitica.* For *Bacillus cereus*, for example, it has been proposed to use a critical level of 10^5 CFUs per gram as a critical level, based on findings from outbreak studies (Carlin *et al.* 2000). With such a critical level, an exposure assessment can be concluded with estimates of the percentage of servings that are above this level, and relative changes in these levels can be applied as a surrogate for relative risks (Nauta *et al.* 2003). A similar approach has been used by Van Damme *et al.* (2017) for *Yersinia enterocolita* in pork, based on hypothetical critical levels.

The use of critical levels is also the basis of the use of the FSO concept (Zwietering *et al.* 2010). The FSO is a critical level that is set on the basis of the acceptable level of protection (ALOP) for the population. It is a useful practical instrument for risk management as it allows validation of the production process (Zwietering *et al.* 2010) and allows the derivation of microbiological criteria (Van Schothorst *et al.* 2009). Still, for risk assessment, it is important to note that, with the biological relevance of single-hit dose–response models, a safe level providing zero risk does not exist, and that the risk is a function of the distribution of doses. The latter implies that even if the level set by the FSO is reached, the risk for different populations in which the FSO is reached may vary considerably.

3.3.2. *Future developments in exposure assessment*

Although the methodology for exposure assessment through the use of food-pathway models and Monte-Carlo simulations is by now well-established, several challenges remain. An important one is that these models require much data, especially the stochastic models that are to be preferred, as explained above. If an exposure assessment is urgently needed, these data demands can frustrate the assessment and more "quick and dirty" methods may be needed.

One important step forward is the development of "knowledge bases", that is, computer systems that combine various sources of models and data in the area of risk assessment and allow the reuse of the developed models. Many of the sometimes complex exposure assessment models need to be reimplemented for re-use in other contexts, which can be resource-demanding and complicate timely risk assessment. Transparency and harmonization of terms and concepts are crucial for such a system (Haberbeck *et al.* 2018; Plaza-Rodriguez *et al.* 2018). The Risk Assessment Modelling and Knowledge Integration Platform (RAKIP) initiative is an important step forward in this respect (Filter *et al.* 2019).

Another important development is the application of whole genome sequencing and data obtained from "omics" technologies in exposure assessment (Den Besten *et al.* 2018; EFSA Panel on Biological Hazards *et al.* 2019; Njage *et al.* 2020). This type of data allows much better characterization of the pathogen in terms of the ability to grow and survive in the food processing conditions, and may therefore provide more precise estimates of the exposure to specific microbial strains. If these characteristics can be linked to virulence factors that are relevant in the hazard characterization, it may allow the identification of "high-risk" strains that can be targeted in hazard identification. Even though these methods may increase the data needs of exposure assessment, they are highly promising.

3.4. References

Carlin, F., Girardin, H., Peck, M.W., Stringer, S.C., Barker, G.B., Martinez, A., Fernandez, A., Fernandez, P., Waites, W.M., Movahedi, S., Van Leusden, F., Nauta, M., Moezelaar, R., del Torre, M., Litman, S. (2000). Research on factors allowing a risk assessment of spore-forming pathogenic bacteria in cooked chilled foods containing vegetables: A FAIR collaborative project. *International Journal of Food Microbiology*, 60, 117–135.

Cassin, M.H., Lammerding, A.M., Todd, E.C.D., Ross, W., McColl, R.S. (1998). Quantitative risk assessment for *Escherichia coli* O157:H7 in ground beef hamburgers. *International Journal of Food Microbiology*, 41, 21–44.

Den Besten, H.M.W., Amezquita, A., Bover-Cid, S., Dagnas, S., Ellouze, M., Guillou, S., Nychas, G., O'Mahony, C., Perez-Rodriguez, F., Membré, J.-M. (2018). Next generation of microbiological risk assessment: Potential of omics data for exposure assessment. *International Journal of Food Microbiology*, 287, 18–27.

EFSA Panel on Biological Hazards, Koutsoumanis, K., Allende, A., Alvarez-Ordonez, A., Bolton, D., Bover-Cid, S., Chemaly, M., Davies, R., De Cesare, A., Hilbert, F., Lindqvist, R., Nauta, M., Peixe, L., Ru, G., Simmons, M., Skandamis, P., Suffredini, E., Jenkins, C., Malorny, B., Ribeiro Duarte, A.S., Torpdahl, M., da Silva Felicio, M.T., Guerra, B., Rossi, M., Herman, L. (2019). Scientific opinion on the whole genome sequencing and metagenomics for outbreak investigation, source attribution and risk assessment of food-borne microorganisms. *EFSA Journal*, 17(12), 5898, 78.

EFSA Scientific Committee, Benford, D., Halldorsson, T., Jeger, M.J., Knutsen, H.K., More, S., Naegeli, H., Noteborn, H., Ockleford, C., Ricci, A., Rychen, G., Schlatter, J.R., Silano, V., Solecki, R., Turck, D., Younes, M., Craig, P., Hart, A., Von Goetz, N., Koutsoumanis, K., Mortensen, A., Ossendorp, B., Martino, L., Merten, C., Mosbach-Schulz, O., Hardy, A. (2018). Guidance on uncertainty analysis in scientific assessments. *EFSA Journal*, 16(1), 5123, 39.

FAO and WHO (2020). Public consultation for draft guidance of microbiological risk assessment for food [Onlne]. Available at: https://www.who.int/news-room/articles-detail/public-consultation-for-draft-guidance-of-microbiological-risk-assessment-for-food [Accessed January 2021].

Filter, M., Rodríguez, C.P., Ungaretti Haberbeck, L., Swaid, A., Georgiadis, S., Valentin, L., Günther, T., Buschhardt, T., Alba Aparicio, M.D., Mesa Varona, O., Desvignes, V., Guillier, L., Nauta, M. (2019). Knowledge platform RAKIP. In *Feed and Food Safety in Times of Global Production and Trade*, Gödert, S. (ed.). Bundesinstitut für Risikobewertung, Berlin [Online]. Available at: https://www.bfr.bund.de/cm/350/feed-and-food-safety-in-times-of-global-production-and-trade.pdf [Accessed January 2021].

Haas, C.N., Rose, J.B., Gerba, C.P. (2014). *Quantitative Microbial Risk Assessment*, 2nd edition. John Wiley & Sons, Hoboken.

Haberbeck, L.U., Plaza-Rodríguez, C., Desvignes, V., Dalgaard, P., Sanaa, M., Guillier, L., Nauta, M., Filter, M. (2018). Harmonized terms, concepts and metadata for microbiological risk assessment models: The basis for knowledge integration and exchange. *Microbial Risk Analysis*, 10, 3–12.

Membré, J.-M. and Valdramidis, V. (2016). *Modeling in Food Microbiology: From Predictive Microbiology to Exposure Assessment.* ISTE Press, London, and Elsevier, Oxford.

Møller, C.O.A., Nauta, M.J., Schaffner, D.W., Dalgaard, P., Christensen, B.B., Hansen, T.B. (2015). Risk assessment of salmonella in Danish meatballs produced in the catering sector. *International Journal of Food Microbiology*, 196, 109–125.

Møller, C.O.A., Nauta, M.J., Hansen, T.B. (2021) Modelling cross-contamination in food processing. In *Risk Assessment Methods for Biological and Chemical Hazards in Food*, Pérez-Rodríguez, F. (ed.). CRC Press, Boca Raton.

Nauta, M.J. (2000). Separation of uncertainty and variability in quantitative microbial risk assessment models. *International Journal of Food Microbiology*, 57, 9–18.

Nauta, M.J. (2001). A modular process risk model structure for quantitative microbiological risk assessment and its application in an exposure assessment of *Bacillus Cereus* in a REPFED. RIVM report 149106007, RIVM Bilthoven [Online]. Available at: https://www.rivm.nl/bibliotheek/rapporten/149106007.pdf [Accessed January 2021].

Nauta, M.J. (2002). Modelling bacterial growth in quantitative risk assessment: Is it possible? *International Journal of Food Microbiology*, 73, 297–304.

Nauta, M.J. (2005). Risk assessment modelling of the food handling processes mixing and partitioning. *International Journal of Food Microbiology*, 100, 311–322.

Nauta, M.J. (2007). Uncertainty and variability in predictive models of microorganisms in food. In *Modelling Microorganisms in Food*, Brul, S., Van Gerwen, S., Zwietering, M. (eds). Woodhead Publishing, Cambridge.

Nauta, M.J. (2008). The modular process risk model (MPRM): A structured approach for food chain exposure assessment. In *Microbial Risk Analysis of Foods*, Schaffner, D.W. (ed.). ASM Press, Washington DC.

Nauta, M.J., Evers, E.G., Takumi, K., Havelaar, A.H. (2001). Risk assessment of Shiga toxin-producing *Escherichia coli* O157 in steak tartare in the Netherlands. RIVM report 257851003, RIVM Bilthoven [Online]. Available at: https://www.rivm.nl/bibliotheek/rapporten/257851003.pdf [Accessed January 2021].

Nauta, M.J., Litman, S., Barker, G.C., Carlin, F. (2003). A retail and consumer phase model for exposure assessment of *Bacillus cereus*. *International Journal of Food Microbiology*, 83(2), 205–218.

Njage, P.M.K., Leekitcharoenphon, P., Hansen, L.T., Hendriksen, R.S., Faes, C., Aerts, M., Hald, T. (2020). Quantitative microbial risk assessment based on whole genome sequencing data: Case of *Listeria monocytogenes*. *Microorganisms*, 8(11), 1772.

Pérez-Rodríguez, F. (2021). *Risk Assessment Methods for Biological and Chemical Hazards in Food*. CRC Press, Boca Raton.

Pérez-Rodríguez, F., Valero, A., Carrasco, E., Garcia, R.M., Zurera, G. (2008). Understanding and modelling bacterial transfer: A review. *Trends in Food Science and Technology*, 19(3), 131–144.

Plaza-Rodríguez, C., Haberbeck, L.U., Desvignes, V., Dalgaard, P., Sanaa, M., Nauta, M., Filter, M., Guillier, L. (2018). Towards transparent and consistent exchange of knowledge for improved microbiological food safety. *Current Opinions in Food Science*, 19, 129–137.

Pouillot, R. and Guillier, L. (2021). Understanding variability and uncertainty in risk assessment. In *Risk Assessment Methods for Biological and Chemical Hazards in Food*, Pérez-Rodríguez, F. (ed.). CRC Press, Boca Raton.

Tenenhaus-Aziza, F. and Ellouze, M. (2015). Software for predictive microbiology and risk assessment: A description and comparison of tools presented at the ICPMF8 Software Fair. *Food Microbiology*, 45, 290–299.

Valdramidis, V., Cummin, E., van Impe, J. (2017). *Quantitative Tools for Sustainable Food and Energy in the Food Chain.* Eurosis Publisher, Ghent.

Van Damme I., De Zutter L., Jacxsens L., Nauta M.J. (2017). Control of human pathogenic *Yersinia enterocolitica* in minced meat: Comparative analysis of different interventions using a risk assessment approach. *Food Microbiology*, 64, 83–95.

Van Schothorst, M., Zwietering, M.H., Ross, T., Buchanan, R.L., Cole, M.B., ICMSF (2009). Relating microbiological criteria to food safety objectives and performance objectives. *Food Control*, 20, 967–979.

Zwietering, M.H. and Nauta, M.J. (2007). Predictive models in food risk assessment. In *Modelling Microorganisms in Food*, Brul, S., Van Gerwen, S., Zwietering, M. (eds). Woodhead Publishing, Cambridge.

Zwietering, M.H., Stewart, C.M., Whiting, R.C., ICMSF (2010). Validation of control measures in a food chain using the FSO concept. *Food Control*, 21, 1716–1722.

PART 2

Food and Risk

4

Pathogenic Mechanisms of Bacterial Foodborne Pathogens

Mayra Aguirre GARCIA and Nabila HADDAD
SECALIM, INRAE, Oniris, Nantes, France

4.1. Introduction

Quantitative microbial risk assessment involves characterizing hazard through its qualitative and/or quantitative evaluation. Hazard characterization includes the determination of the dose–response relationship, that is, the probability and severity of a disease as a function of the dose.

Adverse health effects are influenced by the host's susceptibility and by the nature of the hazard. Foodborne infectious diseases caused by bacterial pathogens can occur as intoxication or infection. In the case of intoxication, a toxin is first produced by bacteria inside the food, and then the contaminated food is ingested by consumers. The toxin pre-formed in the food causes the illness (foodborne poisoning) and constitutes the hazard. In the case of foodborne infection, the pathogen present in the food is the hazard itself. When ingested, it causes disease through toxins and other bacterial effectors produced *in vivo* and/or following the invasion of host tissues.

Once the pathogen has been ingested, a number of outcomes are possible: the pathogen may be excreted or may not survive in the host; it may infect (colonize) but not cause any overt symptoms of illness, resulting in a carrier state for the host. Once the host is infected, there is a probability of acute illness. Once the host falls ill, there is a probability of more severe outcomes,

Microbiological Risk Assessment Associated with the Food Processing and Distribution Chain, coordinated by Jeanne-Marie MEMBRÉ. © ISTE Ltd 2022.

such as septicemia, chronic syndromes and death. Factors that influence these processes include the host's immunological and physiological status and the pathogen's resistance to the host's defense mechanisms, its ability to colonize and its pathogenic mechanisms and virulence factors.

The pathogenicity mechanisms used by the main foodborne pathogens to cause disease are summarized and discussed below.

4.2. Foodborne pathogens and diseases

Foodborne diseases represent important public health issues. Often known as foodborne poisoning or foodborne illnesses, they may be defined as any disease that results from the consumption of food contaminated by bacteria, viruses, parasites or non-infectious agents (e.g. mycotoxins, chemical contaminants). It is worthwhile noting that some foodborne pathogens can contaminate food during all stages by secreting toxins into the extracellular environment, thereby causing poisoning after consumption.

4.2.1. *Definitions of pathogenicity and virulence*

Greater surveillance is not the only way of acquiring a wider approach to foodborne illnesses. We also need to bridge the gaps in the knowledge about the multiple interactions of these pathogens with the environment in their progress along the whole food chain and inside the host during the disease. Pathogenicity and virulence are two terms of central importance in the study of host–pathogen interactions. Nonetheless, these definitions have been extensively discussed and may vary across disciplines. The definitions by Casadevall and Pirofski (2003) and Vega *et al.* (2012) were adopted in this chapter. Pathogenicity is defined as "the disease-producing ability of a microorganism", whereas virulence refers to "the relative capacity of a microorganism to cause damage in a host" (Casadevall and Pirofski 2003; Vega *et al.* 2012). In other words, pathogenicity refers to the ability of an organism to cause disease, while virulence refers to the degree of damage caused by a microbe to its host. In essence, pathogenicity is determined by virulence, which provides the machinery to successfully infect the host cells. Virulence can considerably vary among pathogens, species and inclusive serotypes.

4.2.2. *Epidemiology and clinical manifestations*

Foodborne diseases are thought to be a major cause of morbidity and mortality around the world. A report published by the World Health Organization (WHO) estimated that 31 global hazards caused 600 million foodborne cases and 420,000 deaths in 2010 (WHO 2015; Havelaar *et al.* 2015). However, more than 250 foodborne hazards have been identified (Hoffmann and Scallan 2017), and the real impact of these diseases still seems to be underestimated and is likely unknown (Odeyemi 2016). It is worth noting that this issue cannot be only addressed in developing countries. In the United States, approximately 48 million people are made sick every year by contaminated food (Hoffmann *et al.* 2015). The European Food Safety Authority (EFSA) reported roughly 338,000 confirmed cases attributed to campylobacteriosis and salmonellosis in 2018 (EFSA and ECDC 2019). An analysis within Europe indicates that even though the surveillance system substantially contributes to the monitoring and the reporting of foodborne diseases, the monitoring and reporting are not orchestrated, and reported data mainly include zoonotic infections (Boqvist *et al.* 2018). The magnitude of unsafe food may also be extrapolated to economic costs, with productivity losses related to premature death and post-infectious sequels (Havelaar *et al.* 2015), as well as huge food-safety costs for companies and countries that must meet the global demand and be competitive (Hussain and Dawson 2013). Additionally, antibiotic resistance and increasing food contamination because of environmental changes and dynamic methods of food production threaten to make this problem even more tricky (Newell *et al.* 2010).

Two main categories of foodborne pathogens can be proposed: (i) those that cause the largest number of cases but produce only mild illness and (ii) certain microbes with low incidence but which lead to potentially severe complications requiring frequent hospitalization. According to the WHO's foodborne disease estimation, noroviruses were the most common cause of foodborne illness in Europe in 2010 with approximately 65% of foodborne illnesses (WHO 2017). Nevertheless, foodborne diseases caused by noroviruses have a low mortality rate, and the vast majority of patients experience no major complications. With regard to *Campylobacter* spp., non-typhoidal *Salmonella enterica* and Shiga toxin-producing *Escherichia coli* were among the top five of bacterial pathogens responsible for foodborne diseases in terms of people affected in Europe and the USA (Scallan *et al.* 2011b; WHO 2017; Dewey-Mattia *et al.* 2018). These infectious agents and *Listeria monocytogenes* were estimated to be the leading bacterial causes of hospitalization, fatal cases and disability-adjusted life years (DALYs) resulting from foodborne diseases in Europe in 2010 (WHO 2017; Dewey-Mattia *et al.* 2018). The burden of disease is expressed in DALYs, meaning the years during which a person's life is affected by a disease. Unlike norovirus-related mortality, the mortality rate associated with listeriosis ranges from 25% to 30% (ANSES 2011b). Furthermore, serious complications such as Guillain–Barré syndrome (*Campylobacter jejuni*), hemolytic uremic syndrome (HUS) (*E. coli* STEC) and non-viable pregnancy (*L. monocytogenes*) have been determined. This suggests that although foodborne pathogens generally induce mild to severe enteritis with widely known symptoms, these may differ depending on the different types of foodborne microbes and the health condition of the host. A weakened immune system and age have been established as relevant risk factors.

<table>
<tr><th>Approximate delay before first symptoms occur</th><th colspan="2">Predominant symptoms</th><th>Bacterial pathogen and/or associated toxin</th></tr>
<tr><td>1–7 h, mean 2 h</td><td colspan="2">Nausea, vomiting, retching, diarrhea, abdominal pain, prostration</td><td>Staphylococcus aureus and its enterotoxins</td></tr>
<tr><td>2–4 h</td><td colspan="2">Vomiting, abdominal cramps, nausea</td><td>Bacillus cereus and its emetic toxin</td></tr>
<tr><td>8–16 h</td><td colspan="2">Diarrhea, abdominal cramps, nausea</td><td>Bacillus cereus and its diarrheic toxin</td></tr>
<tr><td>2 h to 6 days, usually 12–36 h</td><td colspan="2">Vertigo, double or blurred vision, loss of light reflex, difficulty swallowing, dry mouth, weakness, respiratory paralysis</td><td>Clostridium botulinum and its neurotoxins</td></tr>
<tr><td>6–96 h (usually 1–3 days)</td><td colspan="2">Fever, abdominal cramps, diarrhea, vomiting, headache</td><td>Salmonella spp., Shigella spp., E. coli</td></tr>
<tr><td>1–10 days (median 3–4 days)</td><td colspan="2">Diarrhea (often bloody), abdominal pain, nausea, vomiting, faintness</td><td>Enterohemorrhagic E. coli</td></tr>
<tr><td>1–8 days (median 2–5 days)</td><td colspan="2">Fever, chills, headache, arthralgia, prostration, dizziness, swollen lymph nodes</td><td>Campylobacter jejuni</td></tr>
<tr><td rowspan="2">Varying periods</td><td>Entire population, all age groups combined</td><td>Septicemia/bacteremia, meningitis, meningoencephalitis, rhombencephalitis, brain abscess, local infections</td><td rowspan="2">Listeria monocytogenes</td></tr>
<tr><td>Pregnant women</td><td>Flu-like symptoms (fever, chills, back pain), spontaneous abortion, death in utero, prematurity, neonatal infection</td></tr>
<tr><td>2–36 h (mean 6–12 h)</td><td colspan="2">Abdominal cramps, diarrhea, putrefactive diarrhea, sometimes nausea and vomiting</td><td>Clostridium perfringens</td></tr>
</table>

Table 4.1. *Symptoms, delay before the first symptoms occur and causal pathogenic bacteria and/or toxins for the main foodborne diseases (adapted from Bintsis (2017))*

4.2.3. *The main bacterial foodborne pathogens*

4.2.3.1. *Bacterial toxin producers in food*

Some foodborne pathogens can drive poisoning via the secretion of harmful molecules such as toxins; they are formed as a result of successful microbial growth in any stage of the food production chain. Following ingestion of food, toxins cause damage in the gastrointestinal epithelium. In some cases, toxins systemically disseminate through the bloodstream or lymphatics, causing damage in other tissues.

For several years, the EFSA has reported more than 5,000 foodborne outbreaks in the European Union (EU) each year (EFSA and ECDC 2018), including illness due to bacterial toxins produced by *Staphylococcus* spp., *Bacillus* spp. and *Clostridium* spp. Even though the number of bacterial toxin-mediated foodborne events is usually underestimated, bacterial toxins are noted to be responsible for nearly 10,000 human cases each year in the EU (EFSA and ECDC 2018). Three toxigenic bacteria are described in this section: *S. aureus*, *B. cereus* and *C. botulinum*.

S. aureus is an important concern in public health because of its diverse arsenal of components that contribute to the pathogenesis of infection, including toxin-mediated virulence, invasiveness and antibiotic resistance. Food spoilage by *S. aureus* enterotoxins is recognized as an important concern for public health and one of the leading causes of foodborne disease worldwide (Kadariya *et al.* 2014). The Center for Disease Control and Prevention (CDC) of the USA estimates that staphylococcal food poisoning causes approximately 241,188 illnesses, 1,064 hospitalizations and 6 deaths in the US each year (Scallan *et al.* 2011b). In 2014, the EU reported 393 foodborne outbreaks caused by staphylococcal toxins (7.5% of all outbreaks), with 2,952 human cases and 264 hospitalizations. Given the fact that humans are the major reservoir for *S. aureus*, food contamination can occur through direct contact or indirectly through skin fragments or respiratory tract droplets. Most staphylococcal food poisoning cases are linked to food contamination during preparation because of inadequate refrigeration, inadequate cooking or heating or poor personal hygiene.

S. aureus is a major human pathogen that can develop a large number of infections. The manifestations of *S. aureus* can range from minor skin problems to more serious infections involving muscle or bone, which may spread to the lung and heart valves (i.e. endocarditis). Worldwide, it is the most common pathogen agent isolated from skin and soft tissue infections and bacteremia, pleuropulmonary, cardiac endothelium, osteoarticular and device-related infections (Tong *et al.* 2015). Toxic shock syndrome, a condition resembling septic shock resulting from the production of toxic shock syndrome toxin 1, has been attributed to *S. aureus* infection (Bacon and Sofos 2003).

With regard to *S. aureus* food poisoning, due to the incubation period of less than 6 and up to 10 h following ingestion of food contaminated with the enterotoxins, the onset of the disease is abrupt. The symptoms may include vomiting, nausea, abdominal cramps, headache, dizziness, chills, perspiration, general weakness, muscle cramping and/or prostration and diarrhea that may or may not contain blood (Bacon and Sofos 2003).

All the enterotoxins produced by *S. aureus* are resistant to digestive enzymes and thermostable. Consequently, they can retain their virulent activity in the digestive tube after ingestion. Similarly, there is no processing treatment compatible with foodstuff consumption, such as pasteurization, ultra-high-temperature treatment or micro-filtration, that can destroy or eliminate the exotoxins produced in food. However, *S. aureus* strains are not all enterotoxic. The percentage of enterotoxinogenic strains varies depending on their origin (Bintsis 2017).

Incidence data related to *B. cereus* food poisoning is particularly limited because of three principal factors. Firstly, it is not a disease for which health authorities must be notified in most countries; secondly, the symptoms of most *B. cereus* infections are generally mild, self-limiting and have a short duration; and thirdly, there is no laboratory testing tool to confirm whether *B. cereus* is the infectious agent responsible for the diseases/symptoms (Enosi Tuipulotu *et al.* 2021). Despite the inaccurate epidemiological surveillance, fatal incidences have been reported (Stenfors Arnesen *et al.* 2008) and foodborne outbreaks associated with *B. cereus* infection demonstrate several trends. In the USA, *B. cereus* is responsible for 0.7% of foodborne illness among the 31 major pathogens (Scallan *et al.* 2011a, 2011b). In the EU, *B. cereus* was implicated in 413 foodborne outbreaks from 2007 to 2014, with 6,657 human cases and 352 hospitalizations in total (EFSA Panel on Biological Hazards 2016).

Most *Bacillus* spp. are found in the environment, including soils and fresh and marine water environments. *B. cereus* spores possess appendages and/or pili and are more hydrophobic than other *Bacillus* spores. Thanks to these properties, the spores adhere to many different types of surfaces and resist removal during cleaning and sanitation (Rajkowski and Smith 2001).

There are two types of foodborne *B. cereus* illnesses: emetic (vomiting), related to food intoxication; and diarrheal, resulting from toxico-infections caused by enteropathogenic strains (Jessberger *et al.* 2020). Diarrheal-type food poisoning results from the action of a thermolabile enterotoxic complex, whereas the emetic-type involves the action of a thermostable toxin. Symptoms are characterized by abdominal pain, nausea and diarrhea following an incubation period of approximately 8–16 h (Table 4.1). Diarrheal syndrome symptoms generally persist for no longer than 12–24 h. Following a 1–5 h incubation period, emetic syndrome symptoms primarily include nausea and vomiting and persist for 6–24 h (Table 4.1). Both disease forms present manifestations that significantly resemble those caused by similar foodborne diseases. The diarrheal and emetic syndromes of *B. cereus* infection may be confused with *Closterium perfringens* and *S. aureus* food poisoning, respectively.

The spore-forming bacteria *C. botulinum* is ubiquitously present in the environment: soil, dust, marine and freshwater sediment, wastewater, slurry and, occasionally, the intestinal tracts of asymptomatic humans and animals (Espelund and Klaveness 2014). The pathogenicity of *C. botulinum* is intimately associated with the production of botulinum neurotoxins, the most acute and lethal bacterial poison currently known.

The botulinum neurotoxin has been traditionally classified into seven serotypes designated with alphabetical letters from A to G (Peck *et al.* 2017). However, more recent molecular analyses, including next-generation sequencing techniques, have led to the discovery of genes encoding for many novel botulinum neurotoxins. They can be added within existing serotypes even if they are composed of different amino acid sequences (Pirazzini *et al.* 2017). Types A, B, E and F cause botulism in humans, types C and D in birds and mammals and type G has yet to be clearly implicated in a botulism case (Peck *et al.* 2017). Similar to many other foodborne diseases, the epidemiological data on botulism is limited due to a low number of reports. In 2014, the EU reported nine foodborne outbreaks caused by *C. botulinum* toxins. The CDC estimate that an average of 27 cases of foodborne botulism occur annually in the USA.

Symptoms of botulinum neurotoxin ingestion appear 12–36 h after the consumption of contaminated food and may initially include nausea and vomiting (Table 4.1). However, these symptoms are followed by more characteristic neurological signs, including visual impairment and acute flaccid paralysis that begins with the muscles of the face, head and pharynx, then moves down to the thorax muscles and extremities, leading to possible death from respiratory failure caused by upper airway or diaphragm paralysis (Bacon and Sofos 2003). The minimum toxic dose of *C. botulinum* neurotoxins has not been determined but, from a human health and food safety standpoint, there should be no tolerance of either the neurotoxin itself or the conditions allowing the growth of the organism in foods (Dewey-Mattia *et al.* 2018).

4.2.3.2. Campylobacter *spp.*

Campylobacter species are well-known as the leading cause of bacterial foodborne diarrheal disease around the world. The WHO has estimated that this pathogen is responsible for approximately 96 million cases annually (WHO 2015). In 2018, the EFSA reported that *Campylobacter* infection was the major cause of zoonosis disease in the EU (EFSA and ECDC 2019). The median disease burden related to *Campylobacter* spp. due to the consumption of foods from an animal source is 27 DALYs per population of 100,000 globally (Li *et al.* 2019), ranking this pathogen in the top three. A total of 99.6 percent of human cases are foodborne and less than 1% are waterborne. In 2018, 25 EU member states reported data concerning *Campylobacter* in food. Like in the previous four years, the highest proportion of units testing positive was observed in fresh meat from broilers (37.5%) (EFSA and ECDC 2019). The highest risk of campylobacteriosis was associated with contaminated broiler products. This was partly due to the fact that all commercial avian species can host *Campylobacter* (Silva *et al.* 2011) and also because of the large quantities of meat consumed in Europe (Humphrey *et al.* 2007). In addition, the most common sources of foodborne outbreaks were milk and broiler meat, similarly to the past years (EFSA and ECDC 2019).

Campylobacter species associated with infections in humans can colonize – often asymptomatically – the lower intestinal tracts of both animals and humans (Dasti *et al.* 2010). In campylobacteriosis cases, the symptoms can range from a self-limiting disease (80% of cases) to permanent and serious extra-intestinal manifestations (Blaser 1997; Dasti *et al.* 2010). Acute illness symptoms include diarrhea (often bloody) in 85% of cases, abdominal pain in 79%, fever in 50%, vomiting in 15% and headaches in 41% (ANSES 2011a). An uncomplicated *Campylobacter* infection lasts between three and seven days for most people (ANSES 2011a). This intestinal disease can be complicated due to dehydration and may require hospitalization.

The genus *Campylobacter* belongs to the family *Campylobacteraceae*. Forty species and subspecies have been isolated from humans and animals. However, *C. jejuni* principally, but also *C. coli*, are considered as the leading causes of foodborne bacterial gastroenteritis worldwide (Igwaran and Okoh 2019). Therefore, research efforts and the insights obtained regarding *Campylobacter* pathogenesis are mainly focused on these two species. Other emerging species are *C. fetus*, *C. lari*, *C. concisus*, *C. upsaliensis* and *C. ureolyticus* (Man 2011; Kaakoush *et al.* 2015).

The *modus operandi* of *Campylobacter* is still poorly understood, probably due to their dissimilar pathogenesis in comparison to other foodborne pathogens (Guerry 2007; Dasti *et al.* 2010). Consequently, the pathogenic mechanisms related to the clinical manifestations still need to be clarified (Bertucci 2015). As of today, the well-documented virulence-associated factors are flagellar motility, bacterial intestinal colonization and invasive and toxin-producing abilities (Dasti *et al.* 2010; Silva *et al.* 2011; Bertucci 2015). In addition, a well-orchestrated plethora of proteins work concomitantly, providing a more hospitable environment for bacterial infection and allowing for pathogen manipulation of host–cell activities (Bertucci 2015). Moreover, other factors, such as genetic tools, may, depending on the strain and the host's immune conditions, alter the infection ability of *Campylobacter* and thereby clinical outcomes. Similarly to other *Campylobacter* features, the pathogenicity of *C. jejuni* has been the main focus of studies and the main knowledge source. Therefore, information about the genus *Campylobacter* has been mainly extrapolated from *C. jejuni* studies.

4.2.3.3. Salmonella enterica

As noted above, *Salmonella* is one of the key causes of foodborne illnesses worldwide. A global systematic analysis recently estimated that enteric fever was responsible for 14.3 million cases, more than 135,000 deaths and 9.8 million DALYs in 2017 (GBD 2017 Typhoid and Paratyphoid Collaborators 2019). Furthermore, non-typhoidal *Salmonella* accounted for approximately 95 million cases of enterocolitis, 59,100 deaths due to non-typhoidal invasive disease and 3.1 million DALYs in the same year (GBD 2017 Causes of Death Collaborators 2018; GBD 2017 Non-Typhoidal *Salmonella* Invasive Disease Collaborators 2019). Children represented a high proportion in both groups. Although the pathogen affects people both in developed and developing countries, lethality in Africa is particularly high, up to 20–25% in some cases (Feasey *et al.* 2012). In addition, *Salmonella* infections can be considered as neglected tropical diseases on this continent (Feasey *et al.* 2012). The main vehicles of transmission are food or polluted feedstuff and water contaminated by contact with fecal matter. Poultry products have been a predominant source of foodborne outbreaks reported by agencies around the globe. For instance, eggs, egg products and broiler meat were the sources of more than 50% of foodborne outbreaks in Europe in 2018 (EFSA and ECDC 2019). This health burden itself highlights the urgent need for further investigations regarding the identification and proper surveillance of potential sources, modes of transmission and host–pathogen interactions in order to implement control and preventive measures.

The genus *Salmonella* belongs to the *Enterobacteriaceae* family and is a facultative, Gram-negative and non-spore forming pathogen. Two species can be broadly distinguished: *S. enterica* and *S. bongori*. However, a third species was accepted in 2005 and identified as *S. subterranean* (Su and Chiu 2007). *S. enterica* causes the vast majority of salmonellosis cases in humans and domestic animals; the species includes six subspecies and more than 2,500 serovars to date (Fabrega and Vila 2013). In order to simplify the classification of diseases caused by *Salmonella* and the respective associated serovars, two types are broadly recognized. On the one hand, *S. enterica* subspecies serovars Typhi and Paratyphi A, B and C are grouped as “enteric fever” and characterized by systemic bacterial infection and common symptoms such as fever, coated tongue, bradycardia, rose spots on the chest and myalgia (Andino and Hanning 2015). On the other hand, non-typhoidal serovars (NTSs) cause self-limiting gastroenteritis or enterocolitis and primarily involve two serovars: *Salmonella* enteritidis and *Salmonella* typhimurium (Bhunia 2018). In addition to these two groups, the *S. enterica* serovars can be divided into three broad classes on the basis of their host specificity (as reviewed by Uzzau *et al.* (2000)). Ubiquitous serovars (e.g. the *S. enterica* serovar Typhimurium) produce acute but self-limiting enteritis in a broad range of hosts, whereas host-specific serovars (e.g. the *S. enterica* serovar Gallinarum) are associated with severe systemic disease in a single species, which may not involve diarrhea. Host-restricted serovars are primarily associated with systemic disease in one host but may cause disease in a limited number of other species (e.g. the *S. enterica* serovar Dublin in cattle and humans).

The main niche of *Salmonella* is the intestinal tract of humans and warm-blooded animals. As a resilient microorganism, *Salmonella* can infect a large number of hosts. This pathogen has an effective genome that enables it to react and evolve in different harmful or stressful environments. Some strains have exhibited a growing resistance to quinolones, fluoroquinones and third-generation cephalosporins, essential antibiotics in the treatment of infection (Threlfall 2002; Andino and Hanning 2015). Likewise, *Salmonella's* molecular machinery enables it to successfully invade human epithelial cells. Some virulence mechanisms have been broadly documented, such as the ability to induce its own phagocytosis in cells, the type-III secretion system (T3SS), virulence structures and the pathogenic factors encoded by its 12 pathogenicity islands (Eng *et al.* 2015; Bhunia 2018).

4.2.3.4. *Diarrheagenic* Escherichia coli

Diarrheal episodes due to diarrheagenic *E. coli* (DEC) infection are an important public health issue for children and adults in developing countries because of their association with morbidity and mortality among children less than 5 years of age. DEC is indeed responsible for about 30–40% of acute diarrhea episodes in children <5 years in developing countries (Miliwebsky *et al.* 2016), and it is an important cause of both sporadic cases and diarrheal outbreaks all over the world (Croxen *et al.* 2013).

In 2018, 8,161 confirmed cases of Shiga toxin-producing *E. coli* (STEC) infection in humans were reported in the EU (EFSA and ECDC 2019). There was strong evidence that the foodborne disease outbreaks associated with STEC in 2018 were mainly due to milk products (cheese, milk), mixed red meat and derived products, vegetables and juices and other derived products (EFSA and ECDC 2019).

Three main groups of *E. coli* strains have been described: commensal or probiotic strains, intestinal-pathogenic strains and extraintestinal-pathogenic strains. This section focuses on intestinal-pathogenic *E. coli*, responsible for gastroenteritis or colitis. Diarrheagenic *E. coli* are classified into six pathotypes according to their preferential host colonization sites, virulence mechanisms and ensuing clinical symptoms and consequences (Table 4.2) (Nataro and Kaper 1998; Kaper *et al.* 2004). We can list enteropathogenic *E. coli* (EPEC), enterohemorrhagic *E. coli* (EHEC), enterotoxigenic *E. coli* (ETEC), enteroaggregative *E. coli* (EAEC), enteroinvasive *E. coli* (EIEC) and diffusely adherent *E. coli* (DAEC) (Nataro and Kaper 1998; Kaper *et al.* 2004).

EPEC infection takes the form of acute diarrhea in infants and is often associated with vomiting (Nataro and Kaper 1998; Kaper *et al.* 2004). These strains are characterized by their attachment and effacement (A/E) lesions.

These are accompanied by the intimate attachment of bacterial cells expressing cell-surface intimin (Eae) with the cognate translocated intimin receptor (Tir) expressed on the surface of eukaryotic cells following injection into their cytosol by a T3SS (Desvaux *et al.* 2006; Nougayrede *et al.* 2006; Lai *et al.* 2013).

EHEC are responsible for pediatric diseases ranging from watery diarrhea to hemorrhagic colitis, which that can further lead to thrombotic microangiopathies, especially HUS and thrombotic thrombocytopenic purpura (TTP) (Ruggenenti and Remuzzi 1998; Tarr *et al.* 2005; Tarr 2009). EHEC are primarily characterized by the expression of the key virulence factor Shiga toxin (Stx), which binds to glycosphingolipid Gb3 before penetrating into the cell and inducing apoptosis of kidney epithelial cells (Melton-Celsa 2014). There are many serotypes of Stx-producing *E. coli*, but only those that have been clinically associated with hemorrhagic colitis are designated as EHEC (Feng *et al.* 2020).

ETEC and EAEC are mainly associated with traveler's diarrhea but can also appear in weanling diarrhea among children (Nataro *et al.* 1998; Jenkins 2018). The diarrhea is generally not bloody but watery, and is sometimes associated with vomiting that ranges from mild to severe purging, similar to cholera.

	ETEC	EPEC	EHEC	EIEC
Infectious dose	High	High	Low	High
Toxin	LT/ST*		Shiga or vero toxins (Stx or VT)	
Invasive				+
Intimin		+	+	
Enterohemolysin			+	
Stool	Watery	Watery, bloody	Watery, very bloody	Mucoid, bloody
Fever	Low	+	-	+
Intestine involved	Small intestine	Small intestine	Colon	Colon, lower small intestine
Serology	Various	026, 0111 and others	O157:H7, O26, O111 and others	Various

Table 4.2. *Some properties and symptoms associated with diarrheagenic* E. coli *pathotypes (according to Feng* et al. *(2020)). *LT, labile toxin; ST, stable toxin*

4.2.3.5. Listeria monocytogenes

According to an EFSA report, the number of people affected by listeriosis in 2018 was similar to the number in 2017 (2,549 cases reported in the EU in 2018 against 2,480 the previous year) (EFSA and ECDC 2019). However, the trend has increased over the past 10 years.

L. monocytogenes is the only *Listeria* species that regularly infects humans, although rare cases of human infections by *L. ivanovii* (a pathogen of ruminants) and *L. grayi* have been reported (Guillet *et al.* 2010; Salimnia *et al.* 2010), and *L. seeligeri* and *L. welshimeri* have been reported as agents of sporadic cases of human listeriosis (Gouin *et al.* 1994). One of three clinical syndromes usually characterizes human listeriosis, namely febrile gastroenteritis, maternal–fetal/neonatal listeriosis or bacteremia with or without brain infection (Vázquez-Boland *et al.* 2001). Febrile gastroenteritis is a self-limited disease characterized by diarrhea, fever, abdominal pain, chills, headache and myalgia (Frye *et al.* 2002), whereas invasive listeriosis leads to more serious infections, such as sepsis, meningitis, endocarditis, encephalitis, meningoencephalitis, septicemia and brain infection. This disease occurs in specific segments of the population – the elderly, pregnant women, unborn babies and immunocompromised people, such as those suffering from acquired immune deficiency syndrome (AIDS) or cancer or those who have undergone an organ transplant (de Noordhout *et al.* 2014).

The success of *L. monocytogenes* in the induction of infection lies in its ability to promote its own internalization into host cells (Carvalho *et al.* 2014). This pathogen can cross three important barriers in the human host, namely the intestinal epithelium, the blood–brain barrier and the placenta, and subsequently disseminate to other organs (Chen *et al.* 2009). The process of infection of host cells by *L. monocytogenes* involves several different stages: adhesion to and invasion of host cells, internalization by host cells, vacuole lysis, intracellular multiplication and intercellular spread to the adjacent cells (Vázquez-Boland *et al.* 2001).

4.3. Pathogenic mechanisms of foodborne pathogens

4.3.1. *Foodborne intoxication: mechanisms of action of toxins pre-formed in food*

Foodborne outbreaks of staphylococcal origin generally only appear from a contamination level in the order of 10^5–10^6 CFUs per gram of food if the strain is an enterotoxin producer. *Staphylococcus aureus* enterotoxin does not directly act on the cells of the intestinal lining but on the nerve endings of the digestive tract. The term "neurotoxin" would therefore be more appropriate. Binding of the toxin to gastrointestinal receptors stimulates the bulbar centers of vomiting (area postrema) through the sensory fibers of the vagus nerve. This mechanism of action explains the reduced incubation time and the uncontrollable nature of vomiting. The amount of toxin needed to cause poisoning varies from 1 to 10 mg per individual, but some sensitive subjects may present disorders from 100 ng.

Although there are nine identified staphylococcal enterotoxins, designated as A, B, C1, C2, C3, D, E, F and G, types A and D are responsible for the majority of outbreaks (Argudín *et al.* 2010). Staphylococcal enterotoxins are included in a larger family of toxins known as pyrogenic toxins, which have the unique ability to act as superantigens that stimulate an extraordinarily high percentage of T cells. They are difficult to inactivate with heat because temperatures required to inactivate them are higher than those needed to kill the organism (Bacon and Sofos 2003).

Another pathogenic mechanism is also considered to explain some of the symptoms observed during poisoning (diarrhea, dizziness, chills, inflammation of the digestive mucosa, etc.). In fact, *S. aureus* enterotoxins are superantigens capable of stimulating large populations of T lymphocytes and lymphokine production.

B. cereus pathogenicity is based on an array of virulence factors that are far from understood and which contribute to a range of human diseases. *B. cereus* can cause two types of food poisoning, primarily characterized by diarrhea and emesis. The emetic syndrome is caused by the emetic toxin produced by the bacteria during their growth phase in the food. The emetic toxin is thermally stable (at 126°C for 90 min, at +4°C for 2 months), even in the presence of trypsin, pepsin or other proteolytic enzymes. The diarrheal syndrome is caused by diarrheal toxins produced in the small intestine during bacterial growth. The rapid onset of the emetic type is characterized by nausea and vomiting, while the late onset of the diarrheal type is characterized by diarrhea and abdominal pain. Both syndromes (diarrheal and emetic) are a result of *B. cereus* endospores surviving the cooking process, after which germination and subsequent proliferation of vegetative cells occur at some point during storage.

Due to its sensitivity to lukewarm temperatures and proteolytic enzymes, the diarrhea toxin is only rarely the cause of intoxication (or disease caused by a toxin present in the food). It is in fact easily destroyed by mere heating of the food and during gastric transit. On the other hand, it can act following the ingestion of vegetative forms of *B. cereus* in the food and the release of the toxin *in vivo* according to a toxi-infectious process.

Botulinum neurotoxin is synthesized during cellular growth and is subsequently released during cell lysis, when proteolytic cleavage activates the molecule (Bacon and Sofos 2003). Four categories of botulism have been described: (i) foodborne botulism derived from ingestion of the toxin present in foods, (ii) wound botulism resulting from toxin production after microorganism growth in an infected wound, (iii) infant botulism from toxin production in the intestinal tract of infants and (iv) botulism due to intestinal colonization in older children and adults with intestinal disorders or complications resulting in a lack of microbial competition. The botulinum neurotoxin introduced in any of these categories is transported via the bloodstream to neuromuscular junctions, where the toxin irreversibly binds to receptors on peripheral nerve endings and is subsequently internalized into the nerve cell (Bacon and Sofos 2003). After ingestion, the toxin is absorbed by the intestine, and carried by the blood and the lymph to the peripheral nervous system. The toxin then acts by blocking the release of neurotransmitters at the cholinergic synapses afferent to the myoneural junction. Disruption of the acetylcholine release mechanism delays or inhibits the conduction of nerve impulses. The mechanism of action of these neurotoxins is as follows:

– binding to a receptor on the surface of the target nerve cells;

– internalization by the target cell via endocytosis during the recycling of acetylcholine;

– translocation through endocytosis vesicles and inhibition of neurotransmitter release.

These toxins are heat-labile. They are destroyed by cooking at 80°C for 10 minutes. However, botulinum toxins are considered as the most virulent bacterial poisons, with a minimum lethal dose of 30 $pg.Kg^{-1}$; 1 mg of this toxin could kill 33 billion mice. *C. botulinum* A toxin is four times more toxic than tetanus toxin and a million times more toxic than strychnine.

4.3.2. *Foodborne infection: pathogenic mechanisms of pathogen colonizers*

The pathogenic mechanism is widely known for including colonization of the digestive tract and adhesion to epithelial intestinal cells, which are eventually followed by invasion of tissues, translocation and toxin production (see Figure 4.1).

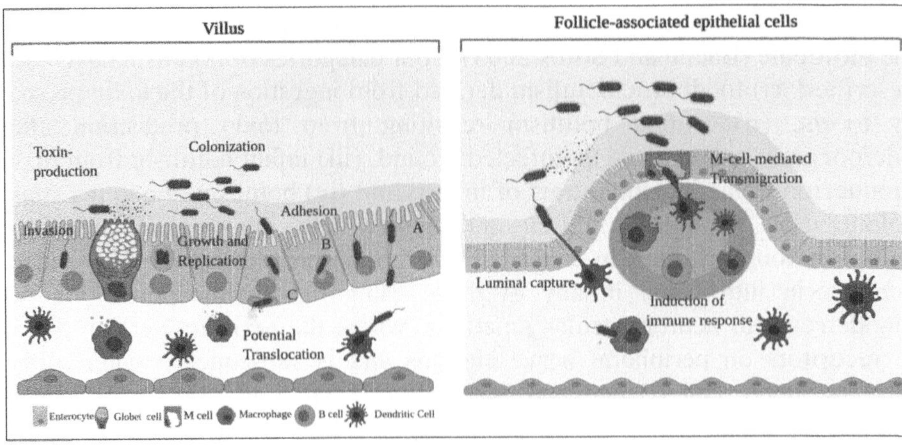

Figure 4.1. *Simplified schematic representation of bacterial virulence events in the intestinal epithelial barrier. Foodborne pathogens can cross the intestinal barrier and spread into the bloodstream. The schematic representation shows the main events of foodborne pathogen virulence, which occur in epithelial cells forming the villus or in follicle-associated epithelial cells. Potential translocation can involve three different mechanisms: A) paracellular, B) transcellular or C) facilitated by vacuoles containing the pathogen. Created with BioRender.com. For a color version of this figure, see www.iste.co.uk/membre/microbiological.zip*

4.3.2.1. *Infectious doses and colonization of the digestive tract*

The determination of the infectious dose of a foodborne pathogen is one of the most important and fundamental steps in evaluating outbreaks of foodborne infections. Only a few studies have evaluated the infectious doses of foodborne pathogens. They concerned *Salmonella*, *E. coli* O157, enterotoxigenic *E. coli* and *C. jejuni*.

Diverse sources provide information about minimum infectious dose data, such as volunteer studies, counts from outbreak investigations and literature reviews related to infectious dose data for a defined population in a particular food matrix. However, these data are inherently uncertain, and variability can be induced by food composition, the health status of the consumer/host and microorganisms (strain dependence). In addition, some pathogens appear to be able to infect their host at low doses of 1–100 units, which are below the detection limit of bacterial detection methods.

Infective doses as low as 500–800 CFUs have been reported for *Campylobacter* spp. (Robinson 1981; Black *et al.* 1988). The amount of ingested EHEC that would cause the disease with high probability was also low. The dose that caused HUS in half of the exposed individuals was assessed to be 600 bacteria for children under the age of 5, and 3 x 10^3 bacteria for children aged 6–10 years. Regarding listeriosis, the estimated infective dose in susceptible populations is 0.1–10 million CFUs versus 10–100 million CFUs in healthy individuals (Angelo *et al.* 2017). Concerning salmonellosis, the human infective dose varies depending on the serovar of the pathogen. Results from volunteer studies indicated that the infective dose for various serovars was 10^5–10^{10} organisms, whereas data from outbreaks suggest that it may be as high as 10^7–10^9 organisms (Taylor *et al.* 1984) or less than 100 organisms (Greenwood and Hooper 1983; Fontaine *et al.* 1980; D'Aoust 1985).

Upon ingestion through contaminated food and in order to reach the site of infection (primarily the intestine), these microbial foodborne agents must successfully overcome the first line of host defenses, mainly composed of saliva, gastric acid, non-specific inflammatory attacks and proteolytic enzymes from the host system, peristalsis and a very dynamic intestinal environment (Green-Johnson 2006; Ponniah *et al.* 2010). The passage through the acidic environment of the stomach can destroy a large proportion of the ingested dose, depending on the buffering capacity of the food. The remaining bacteria can survive and then colonize the lumen of the intestinal tract and multiply inside it.

Once in the intestine, bacteria enter in contact with the intestinal mucus layer, mainly composed of glycoproteins called mucins, digestive enzymes, antimicrobial peptides and immunoglobulins. Although the mucus constitutes a frontline barrier limiting bacterial invasion, some pathogens have evolved mechanisms to go through it and reach epithelial cells. They either produce proteases and directly target host mucins, move on via flagellum-based motility or resist antimicrobial products (Boneca *et al.* 2007; Guerry 2007; Raffatellu *et al.* 2009).

The bacterial flagellum and its components are essential for initial host–pathogen interactions and facilitate intestinal colonization. This important feature presumably enables pathogens to swim towards a variety of substances or away from harmful environments. As an example, *Campylobacter* flagellar motility is driven by a chemotactic signaling system, like the motility of other bacteria (Yao *et al.* 1997; Hartley-Tassell *et al.* 2010). The *C. jejuni* flagellum is composed of O-linked glycosylates and is crucial for crossing the mucus layer and for attachment to the intestinal epithelial cells (Dasti *et al.* 2010). Similarly, pili located on the surface of Gram-negative bacteria enable them to bind to host cell receptors composed of glycolipids, glycoproteins or mucus.

Once in interaction with intestinal epithelial cells, some pathogenic bacteria induce damage by adhering to epithelial intestinal cells and producing toxins *in vivo*, while others can invade host cells and even get access to deeper tissues, promoting their dissemination inside their host.

4.3.2.2. *Toxin production* in vivo

Toxins produced by foodborne pathogens inside the host can target intestinal cells and cause diarrheal symptoms. For instance, *C. jejuni* produces the well-characterized cytolethal distending toxin (CDT). CDT is encoded by three genes designated *cdtA*, *cdtB* and *cdtC*, arranged into an operon that encodes three protein subunits, CdtA, CdtB and CdtC, respectively. CDT is produced by different *Campylobacter* species, including *C. jejuni*, *C. lari*, *C. coli*, *C. fetus* and *C. upsaliensis* (Johnson and Lior 1988; Asakura *et al.* 2007). However, the *cdt* gene cluster is distributed in a species-specific manner (Eyigor *et al.* 1999; Bang *et al.* 2001; Asakura *et al.* 2008). CdtA and CdtC are essential for binding to the host cells and delivering CdtB into them (Figure 4.2). Once CdtB is internalized into the cytoplasm, it is transported to the nucleus via the Golgi apparatus and the endoplasmic reticulum (Heywood *et al.* 2005) (Figure 4.2). Inside the nucleus, CdtB shows activity similar to the enzyme deoxyribonuclease (DNase I) and causes cell-cycle arrest in the G2/M phase. This prevents the cells from entering mitosis and finally leads them to death (Figure 4.2) (Lara-Tejero and Galan 2000; Yamasaki *et al.* 2006; Jinadasa *et al.* 2011). The role of CDT in human campylobacteriosis is still unclear (Abuoun *et al.* 2005).

In 2000, Purdy *et al.* stated that CDT might take part in invasiveness and modulation of the host immune response (Purdy *et al.* 2000). Additional research showed that CDT causes interleukin (IL-) 8 secretion in humans and consequently dendritic cell, macrophage and neutrophil recruitment at the infection site (Hickey *et al.* 2000). In contrast, CDT produced by *C. jejuni* did not promote the synthesis of neutralizing antibodies in chickens. Therefore, this toxin seems to have host-specific recognition antigens (Abuoun *et al.* 2005).

Other toxins produced *in vivo* can target cell types other than intestinal cells, leading to the damage of other organs. In this category, EHEC produce Shiga toxins or Stx, also called Vtx (verotoxins). Two classes of Shiga toxins have been described – Stx1 and Stx2 encoded by the *stx1* and *stx2* genes, respectively. Stx2 is the main toxin involved in the most severe cases of the human disease (Fuller *et al.* 2011), and it shows several variants, suggesting possible host specificity.

Stx production occurs after the adhesion of pathogenic *E. coli* to intestinal cells. These toxins are secreted by the pathogen in the intestine. Even in the absence of canonical Stx receptors on intestinal epithelial cells, luminal Stx facilitates the damage of the intestinal barrier, indirectly via effects on the underlying *lamina propria* (Pradhan *et al.* 2020) or directly because Stx2 damages crypt epithelial cells (while Stx1 does not) (Schüller *et al.* 2004).

Once the epithelial cell layer is damaged, Stx reaches the subepithelial layers of the intestinal wall (Acheson *et al.* 1996), inducing a thrombotic microangiopathy in capillaries and arterioles. Then, the toxin crosses the intestinal epithelium, enters the bloodstream and reaches specific glycolipid receptors (globotrios ceramide 3 (Gb3)) on the surface of endothelial cells (Figure 4.2). Stx is formed of subunit A, which exerts a toxic action and inhibits protein synthesis by cells, and of five B subunits that bind the toxin to the Gb3 receptor host cell (Karmali *et al.* 2010). Portion A of the toxin is internalized by the cell and is transported to the Golgi complex and endoplasmic reticulum. This subunit is N-glycosidase, which acts on the 28S portion of ribosomal RNA and also on the ribosomal subunit, inhibiting protein synthesis and causing cell apoptosis (Figure 4.2) (Endo *et al.* 1988; Karmali *et al.* 2010).

Bound to erythrocytes, neutrophils, platelets or within host blood-cell-derived microvesicles in the blood stream, Stx circulates through the entire body. Subsequently, organ damage outside the gastrointestinal tract develops. Endothelial cells of the kidneys and of the central nervous system are directly targeted by Stx (Richardson *et al.* 1988). This leads to the damage to vascular endothelial cells, especially intestinal, renal and brain cells, which are particularly rich in Gb3 receptors in humans (Vernozy-Rozand 2005). This mechanism is therefore at the origin of clinical manifestations with renal or neurological complications.

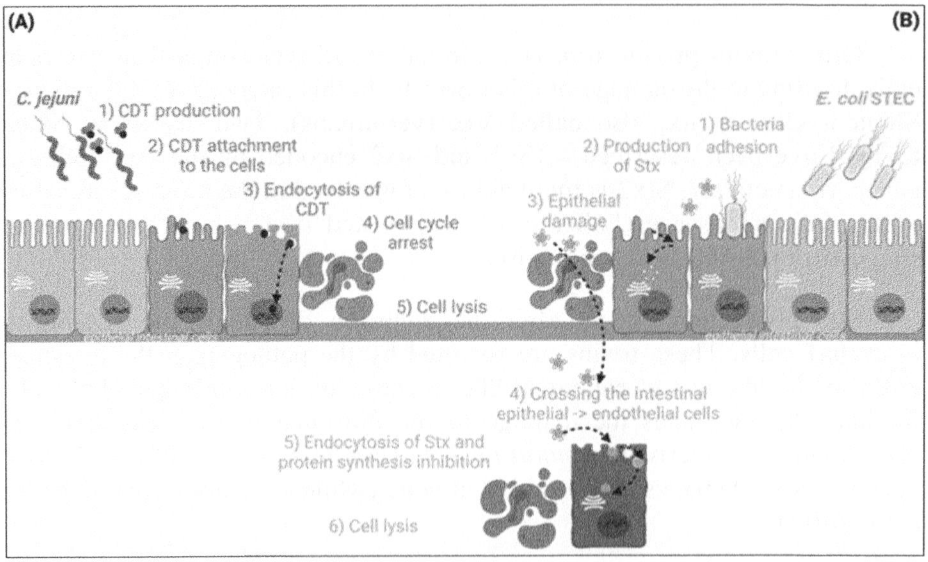

Figure 4.2. *Mechanisms of action of toxins produced* in vivo *by* Campylobacter jejuni *(A) and* E. coli *STEC (B). (A) Production of CDT (A1). Fixation of CdtA and CtdC subunits to the cell (A2), leading to endocytosis of CdtB (B3). CdtB is transported to the nucleus, where it acts as a deoxyribonuclease, leading to DNA fragmentation and cell-cycle arrest (A4). Cell death (A5). (B) Binding of STEC to intestinal cells (B1). Production of Shiga toxin (B2). Epithelial intestinal damage facilitating the translocation of Stx, reaching endothelial cells (B3–B4). Endocytosis of the toxin subunit A into the cell (B5). This subunit is transported to the Golgi complex, and the vesicle is broken. Inhibition of protein synthesis through interaction with 28S rRNA, acting as an N-glycosidase and replacing an adenine moiety. Cell death (B6). Created with BioRender.com*

4.3.2.3. *Adhesion to host cells*

A range of adhesion proteins are involved in the tighter binding of pathogens to the host cells. For example, *L. monocytogenes* internalin A attaches to E-cadherin located in the tight junction of the epithelial barrier. In a comparable manner, *C. jejuni* adhesion factors including fibronectin-binding outer membrane protein CadF bind to fibronectin (Konkel *et al.* 1997). Adhesion to intestinal epithelial cells is a prerequisite for the expression of EHEC pathogenicity. Systems like the T3SS – a needle protein complex – have proved crucial for the adhesion of *E. coli* EHEC (Kenny *et al.* 1997).

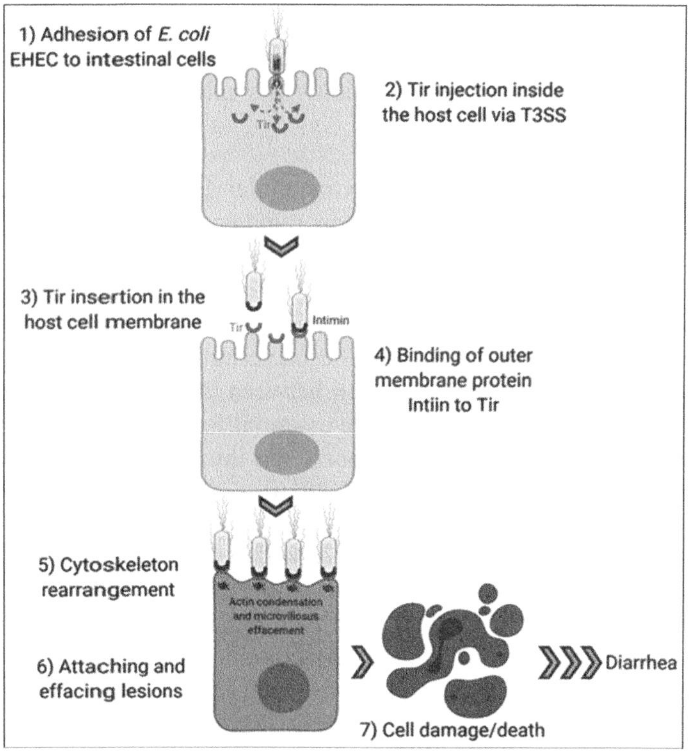

Figure 4.3. *Schematic attachment and effacement mechanisms of* E. coli *EHEC. Infection is initiated by bacterial attachment to the epithelial cells of the small intestine, with subsequent effacement (destruction) of the microvilli (step 1). Following the loose attachment, the active secretion of bacterial proteins into the host epithelial cells occurs through the bacterial type-three secretion system (step 2). The bacterial translocated intimin receptor (Tir) is inserted into the epithelial cell membrane and functions as a receptor for an outer membrane bacterial adhesin – intimin (steps 3 and 4). Binding of intimin to Tir results in actin polymerization, accumulation of cytoskeletal elements beneath the attached bacteria (step 5), loss of cell surface integrity and formation of attaching and effacing lesions, leading to disruption of the brush-border epithelium and causing increased secretion and watery diarrhea (step 6). Cell damage and eventual cell death (step 7). Created with BioRender.com*

EPEC and EHEC use a very particular mechanism to create an intimate contact with host cells: they inject an effector called Tir that inserts itself into the host cell plasma membrane and serves as an "exogenous" receptor for the bacterial surface protein intimin encoded by the *eae* gene (Lai *et al.* 2013). Tir is delivered into the host cell cytoplasm via EPEC or EHEC T3SS, a complex of proteins forming a needle-like structure that crosses the bacterial cell wall and the host cell plasma membrane (Kenny *et al.* 1997). All these proteins and structures are encoded by a set of genes grouped together in an island of four pathogenicity loci called the locus of enterocyte effacement (LEE).

Binding of bacterial intimin to the extracellular domain of Tir is followed by the recruitment of host cell cytoskeleton regulators that locally remodel the actin cytoskeleton (Lai *et al.* 2013). This remodeling leads to the retraction of the host cell absorptive microvilli and to the creation of a pedestal under the attached bacterium, hence the characteristic "attaching and effacing" lesions induced by this pathogen. Tir tethers the bacterium to the host epithelial cell surface and provides a direct connection between the bacterium and the host's cytoskeleton. These bacterial factors are essential for pathogenesis: mutants for the intimin/Tir interaction did not colonize the intestine and were avirulent in animal models of infection (Crepin *et al.* 2010).

The majority of EHEC strains can cause "attachment/effacement" damage to cells in the lining of the distal ileum and colon (Figure 4.3). There exist atypical EHEC strains that do not have the *eae* gene or the LEE. These strains do not induce attachment/effacement lesions but have other mechanisms allowing adhesion to the colonic mucosa. Many potential adhesins have been described, but their involvement in the pathogenic mechanism remains to be established.

4.3.2.4. *Internalization into intestinal epithelial cells*

Adhesion represents a crucial step for extracellular bacteria that facilitates their persistence in the host. For intracellular bacteria, it is a first essential step that precedes their internalization by host cells. An intracellular lifestyle provides diverse advantages to bacterial pathogens: they become inaccessible to humoral and complement-mediated attack and they avoid shear stress-induced clearance and get access to a wide range of nutrients, provided that they can metabolize them. However, host cells also possess different mechanisms specifically targeting these intracellular bacteria. Intracellular pathogens have therefore developed different strategies to successfully establish and maintain intracellular infection.

The invasion of non-phagocytic eukaryotic cells can occur through two main mechanisms that are initially differentiated according to morphological criteria based on the membrane (Figure 4.4). The "trigger" mechanism involves large-scale cytoskeletal rearrangements characterized by the formation of membrane protrusions called ruffles due to the action of bacterial effectors delivered by secretion systems. In contrast, in the "zipper" mechanism, or "receptor-mediated entry", the invading bacteria are tightly bound to the host cell membrane, and only minor cytoskeletal protein rearrangements are initiated by specific contact between bacterial invasins and host cell surface receptors.

After internalization, bacteria can survive and multiply once they are in an intracellular vacuole, such as a lysosome-like vacuoles which have an acidic pH and contain hydrolytic enzymes, or in non-acidic vacuoles that do not fuse to lysosomes and are usually remodeled by the pathogen (Velge *et al.* 2012; Backert *et al.* 2013; Bhunia 2018; Hernandez-Milian and Payeras-Cifre 2014;). Some pathogens can bypass internalization vacuoles and get access to the host cell cytosol, allowing for bacterial release and spread to the bloodstream (Cossart and Roy 2010).

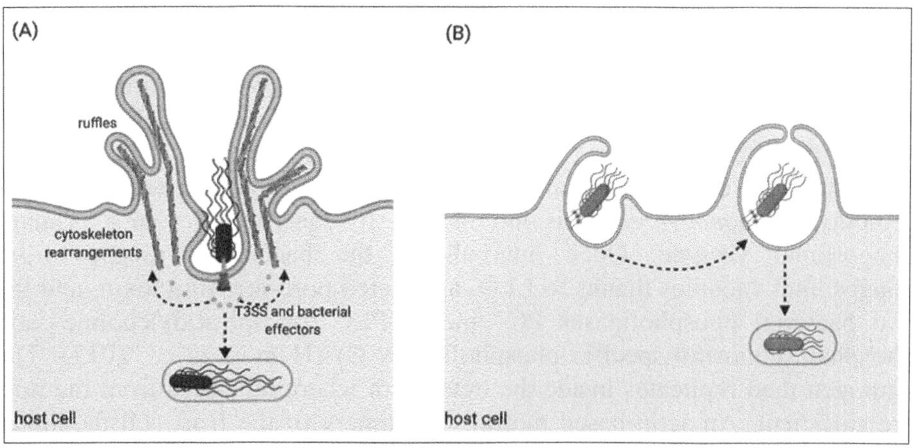

Figure 4.4. *"Zipper" versus "trigger" mechanisms of bacterial internalization by host cells. Schematic representation of pathogen entry inside host cells using a "trigger" (A) or a "zipper" mechanism (B). In the "trigger" mechanism, the injection of effectors by the bacterium into the host cell cytosol provokes important cytoskeletal rearrangements and ruffle formation, allowing for the bacterium to be enfolded and internalized. In the "zipper" mechanism, interactions of bacterial surface proteins with host proteins induce cytoskeleton and membrane rearrangements, leading to bacterial internalization. Created with BioRender.com*

S. enterica induces its internalization into non-phagocytic cells via a trigger mechanism. Using the multi-protein complex T3SS, the pathogen injects at least 15 effectors directly from the bacterial cytoplasm into the host cell cytosol that trigger cellular responses and let the pathogen into the cell (Schlumberger and Hardt 2006; McGhie *et al.* 2009). These effectors modulate cellular processes to the benefit of the pathogen, and five of them (SopE, SopE2, SopB, SipA and SipC) manipulate the cytoskeletal machinery within the host, leading to enfolding. Cell entry is characterized by profuse rearrangements of the actin cytoskeleton at the site of bacterium–host-cell contact that enfold external bacteria and internalize them into membrane-bound vacuoles.

However, evidence shows that *Salmonella* can cause infection in a T3SS-1-independent manner. Two outer membrane proteins – Rck and PagN – have been clearly identified as *Salmonella* invasins. As Rck mediates a zipper-like entry mechanism, *Salmonella* is therefore the first bacterium that has been shown to be able to induce both zipper and trigger mechanisms to invade host cells (Rosselin *et al.* 2010).

In the virulence mechanism of *L. monocytogenes*, the pathogen induces its internalization into non-phagocytic cells via a zipper mechanism (Pizarro-Cerdá and Cossart 2009; Cossart and Helenius 2014). Adhesion and invasion of both phagocytic and non-phagocytic host cells are mediated through surface proteins called internalins (InlA and InlB), which target host plasma membrane proteins (Mengaud *et al.* 1996; Shen *et al.* 2000). After adhering to the epithelial tissue of the gastrointestinal tract through internalin proteins, *L. monocytogenes* is internalized by macrophages in a primary phagosomal vacuole. Once internalized, the bacterium escapes from phagosomal vacuoles thanks to LLO, a secreted pore-forming toxin, and the two bacterial phospholipases PC- and PI-PLC (phosphatidylcholine- and phosphatidylinositol-specific phospholipases C) (Hamon *et al.* 2012). The pathogen then replicates inside the cytoplasm where nutrients from the host are sufficient. An actin-based motility machinery of the host cell facilitates *Listeria* intracellular movement across the cytoplasm to neighboring cells (Neves *et al.* 2013). This intracellular motility leads to the formation of bacterium-containing protrusions and cell-to-cell spread (Vázquez-Boland *et al.* 2001). Upon being internalized by neighboring cells, *L. monocytogenes* is confined in a double-membrane vacuole that it leaves thanks to LLO and *plcB* to restart its life cycle.

Simultaneously to these processes, evasion from the host immune system goes on to cause successful local and systemic infection. Among the mechanisms implemented to elude both innate and adaptive immunities, the manipulation of inflammatory pathways, modulation of the transcription of host immunity genes and interactions with dendritic cells and T and B lymphocytes have been identified (Corr and O'Neill 2009; Bernal-Bayard and Ramos-Morales 2018).

Interestingly, *C. jejuni* internalization may share features of both "zipper" and "trigger" mechanisms (Ó Cróinín and Backert 2012). The invasion ability of *C. jejuni* has been confirmed by their presence in tissue biopsies from patients and different tissue culture models (Grant *et al.* 1993; Rivera-Amill *et al.* 2001; Ó Cróinín and Backert 2012). *C. jejuni* can also invade a number of cultured cell lines *in vitro* and survive there (Ó Cróinín and Backert 2012). Nevertheless, invasion processes can vary significantly, and undermine the efforts made to fully elucidate the factors related to pathogen invasion (Dasti *et al.* 2010). Invasion mechanisms still remain ill-defined. Surprisingly, *C. jejuni* may enter cells with their tip end followed by their flagellar end (Krause-Gruszczynska *et al.* 2007; Dasti *et al.* 2010). One molecular mechanism commonly used by invasive *C. jejuni* strains to enter intestinal epithelial cells is a microtubule-dependent and actin filament-independent invasion mechanism (Monteville 2003; Biswas *et al.* 2003). Additional data also indicate that this mechanism could be involved in the CadF–fibronectin–integrin β1–FAK–paxillin pathway, which is considered one of the major signaling cascades of *C. jejuni*-mediated host cell entry (Ó Cróinín and Backert 2012). Further investigations have revealed other factors, such as phosphatidylinositol 3-kinase (PI3-kinase), protein kinase C (PKC) and a putative mitogen-activated invasion factor, *Campylobacter* invasion antigen B (ciaB). Even though *ciaB* mutant strains have not shown reduced adherence to non-polarized INT-407 cells, a significantly lower number of bacteria have been observed (Konkel *et al.* 1999). However, there is still considerable controversy about the role played by this protein during infection, as one *C. jejuni ciaB* mutant strain was not defective in its invasion ability (Novik *et al.* 2010). Considering intracellular trafficking and survival of *Campylobacter* bacteria, the pathogen seems to localize in a specific compartment in the cytoplasm that would be distinct from lysosomes. It appears that *C. jejuni*-containing vacuoles (CCVs) deviate from the canonical endocytic pathway immediately after host cell entry, thus avoiding delivery into lysosomes (Watson and Galán 2008; Ó Cróinín and Backert 2012).

4.3.2.5. *Translocation through the intestinal barrier*

Several foodborne pathogens, such as *L. monocytogenes*, *Shigella*, *Salmonella* and *Campylobacter*, can transmigrate across the intestinal epithelial barrier, gain access to deeper tissues and cause cell damage and disease in humans (Dasti *et al.* 2010; Ó Cróinín and Backert 2012). Two potential mechanisms related to the passage of pathogens through the intestinal barrier have been described, paracellular and transcellular transmigration routes (Kazmierczak *et al.* 2001; Tegtmeyer *et al.* 2012) (Figure 4.5).

Pathogens that utilize the paracellular mechanism break through the tight junctions and adherent junction complexes and cross the epithelial barrier by passing between neighboring epithelial cells (Figure 4.5). The transcellular mechanism involves the invasion of the pathogen inside epithelial or specialized M cells at the apical surface, followed by intracellular trafficking and exit at the basolateral membrane of host cells (Figure 4.5).

Translocation through non-phagocytic cells of the intestinal epithelium allows pathogens to reach the *lamina propria* and cause systemic infection. Several types of sentinel cells, such as M cells, luminal macrophages and dendritic cells (DCs), continuously sense the presence of pathogenic bacteria in the mucosal environment. Although these cells play a key role in coordinating the innate and adaptative immune responses to limit host colonization by pathogens, they also constitute entry portals for pathogens.

Salmonella may penetrate the intestinal epithelium via a variety of routes. The *S. enterica* serovar Typhimurium invades and destroys M cells located in the follicle-associated epithelium in a T3SS-1-dependent manner, thereby introducing breaches in the intestinal barrier (Jones *et al.* 1994; Penheiter *et al.* 1997). Intestinal villous M cells dispersed in the mucosa and solitary intestinal lymphoid tissue can also take up *Salmonella* (Jang *et al.* 2004; Halle *et al.* 2007). An alternative translocation route involves the T3SS-1-independent transport of bacteria across the epithelium via DCs and/or tissue macrophages (Vazquez-Torres *et al.* 1999). DCs can open tight junctions between epithelial cells, project dendrites into the intestinal lumen to take up *Salmonella* (Rescigno *et al.* 2001; Niess *et al.* 2005) and migrate to distal sites via the blood or the lymph.

In the case of *Listeria*, an interaction between the bacterial InlA surface protein and the host receptor E-cadherin is essential for crossing of the intestinal barrier (Lecuit *et al.* 2001; Bonazzi and Cossart 2011). E-cadherin is a key component of adherens junctions, and was first considered inaccessible for bacteria located in the intestinal lumen. However, accessible E-cadherin is present at sites of cell extrusion at the tip of intestinal villi and at junctions between mucus-secreting goblet cells and adjacent enterocytes. These two locations are sites of invasion by *Listeria* (Pentecost *et al.* 2006; Nikitas *et al.* 2011). In addition to the well-established intracellular lifestyle of *Listeria* involving its escape from the internalization vacuole, its entry through accessible E-cadherin can also lead to its rapid translocation across enterocytes via transcytosis, without escape from the internalization vacuole (Nikitas *et al.* 2011).

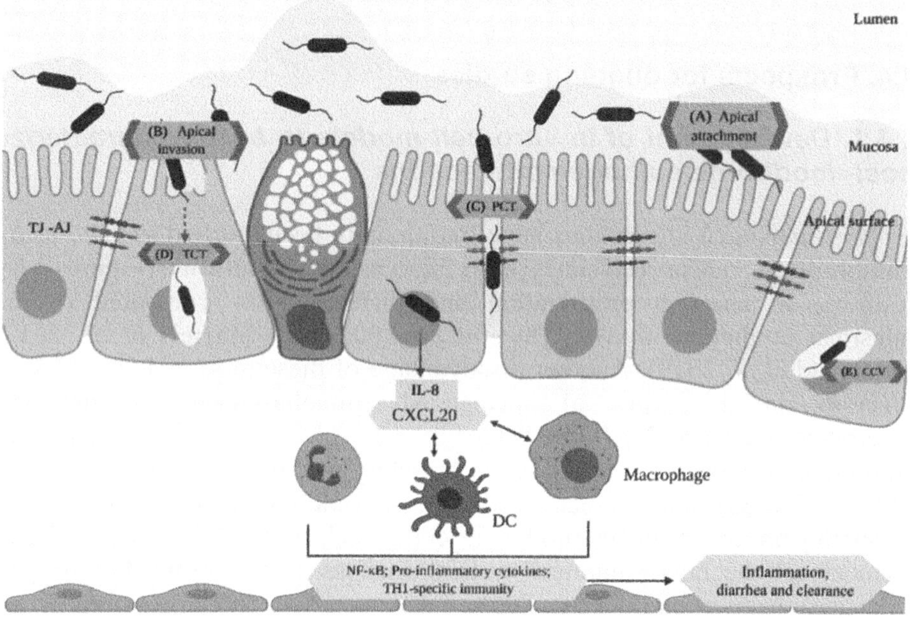

Figure 4.5. *Theoretical* Campylobacter jejuni *mechanisms of intestinal infection.* C. jejuni *adhesion to (A) and invasion of (B) epithelial intestinal cells. Translocation of* C. jejuni *across the intestinal barrier: paracellular (PCT) (C) and transcellular (TCT) (D) mechanisms. Internalization of the pathogen in a* Campylobacter*-containing vacuole (CCV) (E). Abbreviations: TJ–AJ: tight junctions – A junctions; IL-8: interleukin-8; DC: dendritic cell. Created with BioRender.com. For a color version of this figure, see www.iste.co.uk/membre/microbiological.zip*

C. jejuni (Konkel *et al.* 1992; MacCallum *et al.* 2005; Hu *et al.* 2008), *C. coli* (Campana and Baffone 2020) and *C. fetus* (Baker and Graham 2010) can cross intestinal cell monolayers *in vitro* (Backert *et al.* 2013). *C. jejuni* transmigration across polarized cultured cells *in vitro* requires *de novo* protein synthesis and depends on functional flagella. Other studies suggest that bacterial factors such as serine protease high-temperature-resistant protein A (HtrA) (Bæk *et al.* 2011) and lipooligosaccharide (LOS) may participate in bacterial transmigration (Backert *et al.* 2013). However, there is a large gap in our knowledge of the corresponding host cell receptors. The only intensively studied receptor pathway to date, verified by various independent research groups, is the CadF→fibronectin→integrin signaling cascade (Monteville 2003; Krause-Gruszczynska *et al.* 2007; Boehm *et al.* 2012; Eucker and Konkel 2012).

4.4. Prospects for ongoing studies

4.4.1. *Development of* in vitro *cell models to better characterize host–foodborne pathogen interactions*

Host–pathogen interactions have traditionally been studied *in vitro* using immortalized cancer cell lines carrying various mutations (Figure 4.6), resulting in aberrant proliferation and differentiation compared to the intestinal epithelium *in vivo* (Sun *et al.* 2008; von Martels *et al.* 2017; Calatayud *et al.* 2019). Another disadvantage of these models is the lack of diversity in the types of specialized epithelial cells because the enterocyte-like Caco-2 cell line is mostly used. Two-dimensional monolayer co-culture systems have greatly contributed to improving our understanding of host–pathogen interactions. Co-culture or triple culture cell models have frequently involved mucus-producing cells, such as HT29 cells, and basic elements of the innate immune system, such as THP-1 cells (Figure 4.6) (Martinez-Argudo and Jepson 2008; Castiaux *et al.* 2016; Zamora *et al.* 2019; Vernay *et al.* 2020). However, these models still fail to mimic the important features of a native epithelium, making them less suitable to reliably illustrate the pathophysiology of infection. Innovative organoid culture has recently practically overcome the shortcomings of homogeneous cell monolayers (Figure 4.6). Organoids are defined as stem-cell-derived 3D structures with a cell composition comparable to the *in vivo* tissue (Fujii and Sato 2020). The main advantages of the new tissue organoids are their ability to retain their genetic signature and exhibit the complex cellular composition of the original tissues. They can indeed lend themselves to any of the purposes that have been studied with monotypic cell models and more. Consequently, host–foodborne pathogen interaction studies using organoids open onto a new, immensely promising era for addressing novel questions in this research field.

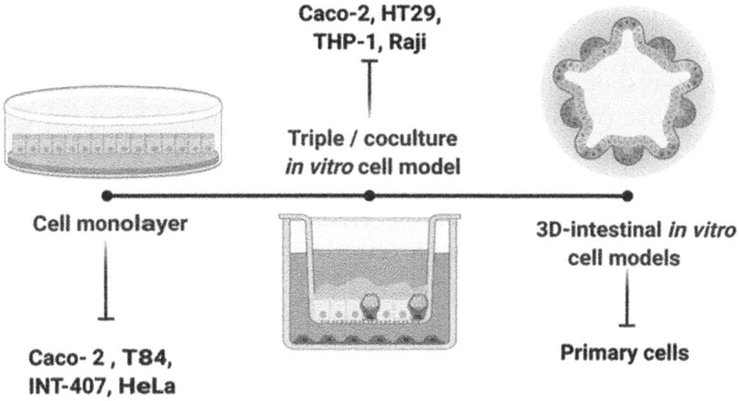

Figure 4.6. *Evolution of* in vitro *cell models to study host–foodborne pathogen interactions. Created with BioRender.com. For a color version of this figure, see www.iste.co.uk/membre/microbiological.zip*

4.4.2. *Better characterization of host–pathogen interactions: towards virulence biomarkers?*

Deciphering host–pathogen interactions by coupling *in vitro* and/or *in vivo* models with -omics technologies, such as genomics, transcriptomics or proteomics, can lead to the identification of bacterial virulence factors and host responses. These data would make it possible to better detect and characterize foodborne pathogens in a food safety context.

If we go further, we might consider biomarkers of virulence or illness by combining quantitative transcriptomic and/or proteomics with virulence and toxicity assays (and/or clinical data) (Haddad *et al.* 2018). Biomarkers related to the disease process or clinical signs could be identified by defining or refining host-related factors linked to pathogen virulence traits. In this case, such biomarkers may help predict clinical outcomes (Haddad *et al.* 2018).

Virulence biomarkers can be relevant for quantitative microbial risk assessment and, more precisely, in hazard characterization, which aims to estimate the probability and severity of a disease as a function of the dose (i.e. the dose–response relationship). Consequently, a precise description of the pathogenic mechanism used by bacteria to cause diseases, and in particular the infectious dose, seems crucial for microbial risk assessment.

If pathogenicity results from a few well-characterized virulence factors are correlated to symptoms, for example, for STEC strains, it may be possible to adjust a dose–response model to the expression of these factors. However, the ingested dose (number of cells) remains relevant. The main challenges here will be to establish and quantify the correlation between the amplitude of the biomarker response and illness. Despite the identification of several virulence factors, their quantification remains insufficient and, until now, no mathematical model has been employed to validate biomarkers.

4.5. References

Abuoun, M., Manning, G., Cawthraw, S.A., Ridley, A., Ahmed, I.H., Wassenaar, T.M., Newell, D.G. (2005). Cytolethal distending toxin (CDT)-negative *Campylobacter jejuni* strains and anti-CDT neutralizing antibodies are induced during human infection but not during colonization in chickens. *Infect Immun*, 73(5), 3053–3062.

Acheson, D.W., Moore, R., De Breucker, S., Lincicome, L., Jacewicz, M., Skutelsky, E., Keusch, G.T. (1996). Translocation of Shiga toxin across polarized intestinal cells in tissue culture. *Infect Immun*, 64(8), 3294.

Andino, A. and Hanning, I. (2015). *Salmonella enterica*: Survival, colonization, and virulence differences among serovars. *ScientificWorldJournal*, 520179.

Angelo, K.M., Conrad, A.R., Saupe, A., Dragoo, H., West, N., Sorenson, A., Barnes, A., Doyle, M., Beal, J., Jackson, K.A., Stroika, S., Tarr, C., Kucerova, Z., Lance, S., Gould, L.H., Wise, M., Jackson, B.R. (2017). Multistate outbreak of *Listeria monocytogenes* infections linked to whole apples used in commercially produced, prepackaged caramel apples: United States, 2014–2015. *Epidemiol Infect*, 145(5), 848–856.

ANSES (2011a). Data sheet on foodborne biological hazards: *Campylobacter Jejuni, Campylobacter coli*. Report, ANSES.

ANSES (2011b). Data sheet on foodborne biological hazards: *Listeria monocytogenes*. Report, ANSES.

Argudín, M.A., Mendoza, M.C., Rodicio, M.R. (2010). Food poisoning and *Staphylococcus aureus* enterotoxins. *Toxins*, 2(7), 1751–1773.

Asakura, M., Samosornsuk, W., Taguchi, M., Kobayashi, K., Misawa, N., Kusumoto, M., Nishimura, K., Matsuhisa, A., Yamasaki, S. (2007). Comparative analysis of cytolethal distending toxin (*cdt*) genes among *Campylobacter jejuni, C. coli* and *C. fetus* strains. *Microb Pathog*, 42(5–6), 174–183.

Asakura, M., Samosornsuk, W., Hinenoya, A., Misawa, N., Nishimura, K., Matsuhisa, A., Yamasaki, S. (2008). Development of a cytolethal distending toxin (*cdt*) gene-based species-specific multiplex PCR assay for the detection and identification of *Campylobacter jejuni*, *Campylobacter coli* and *Campylobacter fetus*. *FEMS Immunol Med Microbiol*, 52(2), 260–266.

Backert, S., Boehm, M., Wessler, S., Tegtmeyer, N. (2013). Transmigration route of *Campylobacter jejuni* across polarized intestinal epithelial cells: Paracellular, transcellular or both? *Cell Communication and Signaling*, 11(1), 72.

Bacon, R.T. and Sofos, J.N. (2003). Characteristics of biological hazards in foods. In *Food Safety Handbook*, Schmidt, R.H. and Rodrick, G.E. (eds). John Wiley & Sons, New Jersey.

Bæk, K.T., Vegge, C.S., Brøndsted, L. (2011). HtrA chaperone activity contributes to host cell binding in *Campylobacter jejuni*. *Gut Pathog*, 3(1), 13.

Baker, N.T. and Graham, L.L. (2010). *Campylobacter fetus* translocation across Caco-2 cell monolayers. *Microb Pathog*, 49(5), 260–272.

Bang, D.D., Scheutz, F., Ahrens, P., Pedersen, K., Blom, J., Madsen, M. (2001). Prevalence of cytolethal distending toxin (*cdt*) genes and CDT production in *Campylobacter* spp. isolated from Danish broilers. *J Med Microbiol*, 50(12), 1087–1094.

Bernal-Bayard, J. and Ramos-Morales, F. (2018). Molecular mechanisms used by *Salmonella* to evade the immune system. *Curr Issues Mol Biol*, 133–168.

Bertucci, B.A. (2015). *Campylobacter Infections: Epidemiology, Clinical Management and Prevention*. Nova Science Publishers, New York.

Bhunia, A.K. (2018). *Foodborne Microbial Pathogens Mechanisms and Pathogenesis*. Springer-Verlag, New York.

Bintsis, T. (2017). Foodborne pathogens. *AIMS Microbiol*, 3(3), 529–563.

Biswas, D., Itoh, K., Sasakawa, C. (2003). Role of microfilaments and microtubules in the invasion of INT-407 cells by *Campylobacter jejuni*. *Microbiol Immunol*, 47(6), 469–473.

Black, R.E., Levine, M.M., Clements, M.L., Hughes, T.P., Blaser, M.J. (1988). Experimental *Campylobacter jejuni* infection in humans. *J Infect Dis*, 157(3), 472–479.

Blaser, M.J. (1997). Epidemiologic and clinical features of *Campylobacter jejuni* infections. *J Infect Dis*, 176(Suppl 2), S103–S105.

Boehm, M., Hoy, B., Rohde, M., Tegtmeyer, N., Bæk, K.T., Oyarzabal, O.A., Brøndsted, L., Wessler, S., Backert, S. (2012). Rapid paracellular transmigration of *Campylobacter jejuni* across polarized epithelial cells without affecting TER: Role of proteolytic-active HtrA cleaving E-cadherin but not fibronectin. *Gut Pathog*, 4(1), 3.

Bonazzi, M. and Cossart, P. (2011). Impenetrable barriers or entry portals? The role of cell–cell adhesion during infection. *J Cell Biol*, 195(3), 349–358.

Boneca, I.G., Dussurget, O., Cabanes, D., Nahori, M.-A., Sousa, S., Lecuit, M., Psylinakis, E., Bouriotis, V., Hugot, J.-P., Giovannini, M., Coyle, A., Bertin, J., Namane, A., Rousselle, J.-C., Cayet, N., Prévost, M.-C., Balloy, V., Chignard, M., Philpott, D.J., Cossart, P., Girardin, S.E. (2007). A critical role for peptidoglycan N-deacetylation in *Listeria* evasion from the host innate immune system. *Pro Nat Acad of Sci*, 104(3), 997–1002.

Boqvist, S., Soderqvist, K., Vagsholm, I. (2018). Food safety challenges and one health within Europe. *Acta Vet Scand*, 60(1), 1.

Calatayud, M., Dezutter, O., Hernandez-Sanabria, E., Hidalgo-Martinez, S., Meysman, F.J.R., Van De Wiele, T. (2019). Development of a host–microbiome model of the small intestine. *FASEB J*, 33(3), 3985–3996.

Campana, R. and Baffone, W. (2020). Intracellular survival and translocation ability of human and avian *Campylobacter jejuni* and *Campylobacter coli* strains. *Adv Exp Med Biol*, 1282, 115–125.

Carvalho, F., Sousa, S., Cabanes, D. (2014). How *Listeria monocytogenes* organizes its surface for virulence. *Front Cell Infect Microbiol*, 4, 48.

Casadevall, A. and Pirofski, L.A. (2003). The damage-response framework of microbial pathogenesis. *Nat Rev Microbiol*, 1(1), 17–24.

Castiaux, V., Laloux, L., Schneider, Y.J., Mahillon, J. (2016). Screening of cytotoxic *B. cereus* on differentiated Caco-2 cells and in co-culture with mucus-secreting (HT29-MTX) cells. *Toxins (Basel)*, 8(11).

Chen, J., Luo, X., Jiang, L., Jin, P., Wei, W., Liu, D., Fang, W. (2009). Molecular characteristics and virulence potential of *Listeria monocytogenes* isolates from Chinese food systems. *Food Microbiol*, 26(1), 103–111.

Corr, S.C. and O'Neill, L.A.J. (2009). *Listeria monocytogenes* infection in the face of innate immunity. *Cell Microbiol*, 11(5), 703–709.

Cossart, P. and Helenius, A. (2014). Endocytosis of viruses and bacteria. *Cold Spring Harb Perspect Biol*, 6(8).

Cossart, P. and Roy, C.R. (2010). Manipulation of host membrane machinery by bacterial pathogens. *Curr Opin in Cell Biol*, 22(4), 547–554.

Crepin, V.F., Girard, F., Schüller, S., Phillips, A.D., Mousnier, A., Frankel, G. (2010). Dissecting the role of the Tir:Nck and Tir:IRTKS/IRSp53 signalling pathways *in vivo*. *Mol Microbiol*, 75(2), 308–323.

Croxen, M.A., Law, R.J., Scholz, R., Keeney, K.M., Wlodarska, M., Finlay, B.B. (2013). Recent advances in understanding enteric pathogenic *Escherichia coli*. *Clin Microbiol Rev*, 26(4), 822–880.

D'Aoust, J.Y. (1985). Infective dose of *Salmonella* Typhimurium in cheddar cheese. *Am J Epidemiol*, 122(4), 717–720.

Dasti, J.I., Tareen, A.M., Lugert, R., Zautner, A.E., Gross, U. (2010). *Campylobacter jejuni*: A brief overview on pathogenicity-associated factors and disease-mediating mechanisms. *Int J Med Microbiol*, 300(4), 205–211.

De Noordhout, C.M., Devleesschauwer, B., Angulo, F.J., Verbeke, G., Haagsma, J., Kirk, M., Havelaar, A., Speybroeck, N. (2014). The global burden of listeriosis: A systematic review and meta-analysis. *Lancet Infect Dis*, 14(11), 1073–1082.

Desvaux, M., Hebraud, M., Henderson, I.R., Pallen, M.J. (2006). Type III secretion: What's in a name? *Trends Microbiol*, 14(4), 157–160.

Dewey-Mattia, D., Manikonda, K., Hall, A.J., Wise, M.E., Crowe, S.J. (2018). Surveillance for foodborne disease outbreaks – United States, 2009–2015. *MMWR Surveill Summ*, 67(10), 1–11.

EFSA and ECDC (2018). The European Union summary report on trends and sources of zoonoses, zoonotic agents and food-borne outbreaks in 2017. *EFSA J*, 16(12), e05500.

EFSA and ECDC (2019). The European Union One Health 2018 zoonoses report. *EFSA J*, 17(12), 1–276.

EFSA Panel on Biological Hazards (2016). Risks for public health related to the presence of *Bacillus cereus* and other *Bacillus* spp. including *Bacillus thuringiensis* in foodstuffs. *EFSA J*, 14(7), e04524.

Endo, Y., Tsurugi, K., Yutsudo, T., Takeda, Y., Ogasawara, T., Igarashi, K. (1988). Site of action of a Vero toxin (VT2) from *Escherichia coli* O157:H7 and of Shiga toxin on eukaryotic ribosomes. RNA N-glycosidase activity of the toxins. *Eur J Biochem*, 171(1–2), 45–50.

Eng, S.K., Pusparajah, P., Ab Mutalib, N.S., Ser, H.L., Chan, K.G., Lee, L.H. (2015). *Salmonella*: A review on pathogenesis, epidemiology and antibiotic resistance. *Front Life Sci*, 8(3), 284–293.

Enosi Tuipulotu, D., Mathur, A., Ngo, C., Man, S.M. (2021). *Bacillus cereus*: Epidemiology, virulence factors, and host-pathogen Interactions. *Trends Microbiol*, 29(5), 458–471.

Espelund, M. and Klaveness, D. (2014). Botulism outbreaks in natural environments – An update. *Front Microbiol*, 5(287), 1–7.

Eucker, T.P. and Konkel, M.E. (2012). The cooperative action of bacterial fibronectin-binding proteins and secreted proteins promote maximal *Campylobacter jejuni* invasion of host cells by stimulating membrane ruffling. *Cell Microbiol*, 14(2), 226–238.

Eyigor, A., Dawson, K.A., Langlois, B.E., Pickett, C.L. (1999). Cytolethal distending toxin genes in *Campylobacter jejuni* and *Campylobacter coli* isolates: Detection and analysis by PCR. *J Clin Microbiol*, 37(5), 1646–1650.

Fabrega, A. and Vila, J. (2013). *Salmonella enterica* serovar Typhimurium skills to succeed in the host: Virulence and regulation. *Clin Microbiol Rev*, 26(2), 308–341.

Feasey, N.A., Dougan, G., Kingsley, R.A., Heyderman, R.S., Gordon, M.A. (2012). Invasive non-typhoidal *Salmonella* disease: An emerging and neglected tropical disease in Africa. *Lancet*, 379(9835), 2489–2499.

Feng, P., Weagant, S.D., Jinneman, K. (2020). Diarrheagenic *Escherichia* coli. In *Bacteriological Analytical Manual*, 8th edition, revision A, FDA (ed.). U.S. Food and Drug Administration.

Fontaine, R.E., Cohen, M.L., Martin, W.T., Vernon, T.M. (1980). Epidemic salmonellosis from cheddar cheese: Surveillance and prevention. *Am J Epidemiol*, 111(2), 247–253.

Frye, D.M., Zweig, R., Sturgeon, J., Tormey, M., Lecavalier, M., Lee, I., Lawani, L., Mascola, L. (2002). An outbreak of febrile gastroenteritis associated with delicatessen meat contaminated with *Listeria monocytogenes*. *Clin Infect Dis*, 35(8), 943–949.

Fujii, M. and Sato, T. (2020). Somatic cell-derived organoids as prototypes of human epithelial tissues and diseases. *Nat Mater*, 20, 156–169.

Fuller, C.A., Pellino, C.A., Flagler, M.J., Strasser, J.E., Weiss, A.A. (2011). Shiga toxin subtypes display dramatic differences in potency. *Infect Immun*, 79, 1329–1337.

GBD 2017 Causes of Death Collaborators (2018). Global, regional, and national age-sex-specific mortality for 282 causes of death in 195 countries and territories, 1980–2017: A systematic analysis for the Global Burden of Disease Study 2017. *Lancet*, 392(10159), 1736–1788.

GBD 2017 Non-Typhoidal *Salmonella* Invasive Disease Collaborators (2019). The global burden of non-typhoidal *Salmonella* invasive disease: A systematic analysis for the Global Burden of Disease Study 2017. *Lancet Infect Dis*, 19(12), 1312–1324.

GBD 2017 Typhoid and Paratyphoid Collaborators (2019). The global burden of typhoid and paratyphoid fevers: A systematic analysis for the global burden of disease study 2017. *Lancet Infect Dis*, 19(4), 369–381.

Gouin, E., Mengaud, J., Cossart, P. (1994). The virulence gene cluster of *Listeria monocytogenes* is also present in *Listeria ivanovii*, an animal pathogen, and *Listeria seeligeri*, a nonpathogenic species. *Infect Immun*, 62(8), 3550–3553.

Grant, C.C., Konkel, M.E., Cieplak, W., Tompkins, L.S. (1993). Role of flagella in adherence, internalization, and translocation of *Campylobacter jejuni* in nonpolarized and polarized epithelial cell cultures. *Infect Immun*, 61(5), 1764–1771.

Green-Johnson, J.M. (2006). Nonspecific host defenses against foodborne pathogens. In *Food Consumption and Disease Risk*, Potter, M. (ed.). Elsevier, Amsterdam.

Greenwood, M.H. and Hooper, W.L. (1983). Chocolate bars contaminated with *Salmonella* Napoli: An infectivity study. *Br Med J (Clin Res Ed)*, 286(6375), 1394.

Guerry, P. (2007). *Campylobacter* flagella: Not just for motility. *Trends Microbiol*, 15(10), 456–461.

Guillet, C., Join-Lambert, O., Le Monnier, A., Leclercq, A., Mechaï, F., Mamzer-Bruneel, M.F., Bielecka, M.K., Scortti, M., Disson, O., Berche, P., Vazquez-Boland, J., Lortholary, O., Lecuit, M. (2010). Human listeriosis caused by *Listeria ivanovii*. *Emerg Infect Dis*, 16(1), 136–138.

Haddad, N., Johnson, N., Kathariou, S., Métris, A., Phister, T., Pielaat, A., Tassou, C., Wells-Bennik, M.H.J., Zwietering, M.H. (2018). Next generation microbiological risk assessment – Potential of omics data for hazard characterisation. *Int J Food Microbiol*, 287, 28–39.

Halle, S., Bumann, D., Herbrand, H., Willer, Y., Dähne, S., Förster, R., Pabst, O. (2007). Solitary intestinal lymphoid tissue provides a productive port of entry for *Salmonella enterica* serovar Typhimurium. *Infect Immun*, 75(4), 1577–1585.

Hamon, M.A., Ribet, D., Stavru, F., Cossart, P. (2012). Listeriolysin O: The Swiss army knife of *Listeria*. *Trends Microbiol*, 20(8), 360–368.

Hartley-Tassell, L.E., Shewell, L.K., Day, C.J., Wilson, J.C., Sandhu, R., Ketley, J.M., Korolik, V. (2010). Identification and characterization of the aspartate chemosensory receptor of *Campylobacter jejuni*. *Mol Microbiol*, 75(3), 710–730.

Havelaar, A.H., Kirk, M.D., Torgerson, P.R., Gibb, H.J., Hald, T., Lake, R.J., Praet, N., Bellinger, D.C., De Silva, N.R., Gargouri, N., Speybroeck, N., Cawthorne, A., Mathers, C., Stein, C., Angulo, F.J., Devleesschauwer, B., World Health Organization Foodborne Disease Burden Epidemiology Reference Group (2015). World Health Organization global estimates and regional comparisons of the burden of foodborne disease in 2010. *PLoS Med*, 12(12), e1001923.

Hernandez-Milian, A. and Payeras-Cifre, A. (2014). What is new in listeriosis? *BioMed Res Int*, 2014, 1–7.

Heywood, W., Henderson, B., Nair, S.P. (2005). Cytolethal distending toxin: Creating a gap in the cell cycle. *J Med Microbiol*, 54(Pt 3), 207–216.

Hickey, T.E., McVeigh, A.L., Scott, D.A., Michielutti, R.E., Bixby, A., Carroll, S.A., Bourgeois, A.L., Guerry, P. (2000). *Campylobacter jejuni* cytolethal distending toxin mediates release of interleukin-8 from intestinal epithelial cells. *Infect Immun*, 68(12), 6535–6541.

Hoffmann, S. and Scallan, E. (2017). Epidemiology, cost, and risk analysis of foodborne disease. In *Foodborne Diseases*, Dodd, C.E.R., Aldsworth, T., Stein, R.A., Cliver, D.O., Riemann, H.P. (eds). Elsevier, Amsterdam.

Hoffmann, S., Maculloch, B., Batz, M. (2015). Economic burden of major foodborne illnesses acquired in the United States. Economic Information Bulletin, US Department of Agriculture.

Hu, L., Tall, B.D., Curtis, S.K., Kopecko, D.J. (2008). Enhanced microscopic definition of *Campylobacter jejuni* 81–176 adherence to, invasion of, translocation across, and exocytosis from polarized human intestinal Caco-2 cells. *Infect Immun*, 76(11), 5294–5304.

Humphrey, T., O'Brien, S., Madsen, M. (2007). Campylobacters as zoonotic pathogens: A food production perspective. *Int J Food Microbiol*, 117(3), 237–257.

Hussain, M.A. and Dawson, C.O. (2013). Economic impact of food safety outbreaks on food businesses. *Foods*, 2(4), 585–589.

Igwaran, A. and Okoh, A.I. (2019). Human campylobacteriosis: A public health concern of global importance. *Heliyon*, 5(11), e02814.

Jang, M.H., Kweon, M.-N., Iwatani, K., Yamamoto, M., Terahara, K., Sasakawa, C., Suzuki, T., Nochi, T., Yokota, Y., Rennert, P.D., Hiroi, T., Tamagawa, H., Iijima, H., Kunisawa, J., Yuki, Y., Kiyono, H. (2004). Intestinal villous M cells: An antigen entry site in the mucosal epithelium. *Pro Nat Acad Sci*, 101(16), 6110–6115.

Jenkins, C. (2018). Enteroaggregative *Escherichia coli*. *Curr Top Microbiol Immunol*, 416, 27–50.

Jessberger, N., Dietrich, R., Granum, P.E., Märtlbauer, E. (2020). The *Bacillus cereus* food infection as multifactorial process. *Toxins*, 12(11), 701.

Jinadasa, R.N., Bloom, S.E., Weiss, R.S., Duhamel, G.E. (2011). Cytolethal distending toxin: A conserved bacterial genotoxin that blocks cell cycle progression, leading to apoptosis of a broad range of mammalian cell lineages. *Microbiology (Reading)*, 157(Pt 7), 1851–1875.

Johnson, W.M. and Lior, H. (1988). A new heat-labile cytolethal distending toxin (CLDT) produced by *Campylobacter* spp. *Microb Pathog*, 4, 115–126.

Jones, B.D., Ghori, N., Falkow, S. (1994). *Salmonella* typhimurium initiates murine infection by penetrating and destroying the specialized epithelial M cells of the Peyer's patches. *J Exp Med*, 180(1), 15–23.

Kaakoush, N.O., Castano-Rodriguez, N., Mitchell, H.M., Man, S.M. (2015). Global epidemiology of *Campylobacter* infection. *Clin Microbiol Rev*, 28(3), 687–720.

Kadariya, J., Smith, T.C., Thapaliya, D. (2014). *Staphylococcus aureus* and staphylococcal food-borne disease: An ongoing challenge in public health. *BioMed Res Int*, 2014, 827965.

Kaper, J.B., Nataro, J.P., Mobley, H.L.T. (2004). Pathogenic *Escherichia coli*. *Nat Rev Microbiol*, 2, 123–140.

Karmali, M.A., Gannon, V., Sargeant, J.M. (2010). Verocytotoxin-producing *Escherichia coli* (VTEC). *Vet Microbiol*, 140(3–4), 360–370.

Kazmierczak, B.I., Mostov, K., Engel, J.N. (2001). Interaction of bacterial pathogens with polarized epithelium. *Annu Rev Microbiol*, 55(1), 407–435.

Kenny, B., Devinney, R., Stein, M., Reinscheid, D.J., Frey, E.A., Finlay, B.B. (1997). Enteropathogenic *E. coli* (EPEC) transfers its receptor for intimate adherence into mammalian cells. *Cell*, 91(4), 511–520.

Konkel, M.E., Mead, D.J., Hayes, S.F., Cieplak, W. (1992). Translocation of *Campylobacter jejuni* across human polarized epithelial cell monolayer cultures. *J Infect Diseases*, 166(2), 308–315.

Konkel, M.E., Garvis, S.G., Tipton, S.L., Anderson, J.D.E., Cieplak, J.W. (1997). Identification and molecular cloning of a gene encoding a fibronectin-binding protein (CadF) from *Campylobacter jejuni*. *Mol Microbiol*, 24(5), 953–963.

Konkel, M.E., Kim, B.J., Rivera-Amill, V., Garvis, S.G. (1999). Bacterial secreted proteins are required for the internalization of *Campylobacter jejuni* into cultured mammalian cells. *Mol Microbiol*, 32(4), 691–701.

Krause-Gruszczynska, M., Rohde, M., Hartig, R., Genth, H., Schmidt, G., Keo, T., König, W., Miller, W.G., Konkel, M.E., Backert, S. (2007). Role of the small Rho GTPases Rac1 and Cdc42 in host cell invasion of *Campylobacter jejuni*. *Cell Microbiol*, 9(10), 2431–2444.

Lai, Y., Rosenshine, I., Leong, J.M., Frankel, G. (2013). Intimate host attachment: Enteropathogenic and enterohaemorrhagic *Escherichia coli*. *Cell Microbiol*, 15(11), 1796–1808.

Lara-Tejero, M. and Galan, J.E. (2000). A bacterial toxin that controls cell cycle progression as a deoxyribonuclease I-like protein. *Science*, 290(5490), 354–357.

Lecuit, M., Vandormael-Pournin, S., Lefort, J., Huerre, M., Gounon, P., Dupuy, C., Babinet, C., Cossart, P. (2001). A transgenic model for listeriosis: Role of internalin in crossing the intestinal barrier. *Science*, 292(5522), 1722–1725.

Li, M., Havelaar, A.H., Hoffmann, S., Hald, T., Kirk, M.D., Torgerson, P.R., Devleesschauwer, B. (2019). Global disease burden of pathogens in animal source foods, 2010. *PLoS One*, 14(6), e0216545.

MacCallum, A., Hardy, S.P., Everest, P.H. (2005). *Campylobacter jejuni* inhibits the absorptive transport functions of Caco-2 cells and disrupts cellular tight junctions. *Microbiology*, 151(7), 2451–2458.

Man, S.M. (2011). The clinical importance of emerging *Campylobacter* species. *Nat Rev Gastroenterol Hepatol*, 8(12), 669–685.

Martinez-Argudo, I. and Jepson, M.A. (2008). *Salmonella* translocates across an *in vitro* M cell model independently of SPI-1 and SPI-2. *Microbiology*, 154(12), 3887–3894.

McGhie, E.J., Brawn, L.C., Hume, P.J., Humphreys, D., Koronakis, V. (2009). *Salmonella* takes control: Effector-driven manipulation of the host. *Curr Opin Microbiol*, 12(1), 117–124.

Melton-Celsa, A.R. (2014). Shiga toxin (Stx) classification, structure, and function. *Microbiol Spectr*, 2(4), EHEC-0024-2013.

Mengaud, J., Ohayon, H., Gounon, P., Mège, R.-M., Cossart, P. (1996). E-cadherin is the receptor for internalin, a surface protein required for entry of *L. monocytogenes* into epithelial cells. *Cell*, 84(6), 923–932.

Miliwebsky, E., Schelotto, F., Varela, G., Luz, D., Chinen, I., Piazza, R.M.F. (2016). Human diarrheal infections: Diagnosis of diarrheagenic *Escherichia coli* pathotypes. In Escherichia coli *in the Americas*, Torres, A.G. (ed.). Springer International Publishing, Basel.

Monteville, M.R. (2003). Maximal adherence and invasion of INT 407 cells by *Campylobacter jejuni* requires the CadF outer-membrane protein and microfilament reorganization. *Microbiology*, 149(1), 153–165.

Nataro, J.P. and Kaper, J.B. (1998). Diarrheagenic *Escherichia coli. Clin Microbiol Rev*, 11(1), 142–201.

Nataro, J.P., Steiner, T., Guerrant, R.L. (1998). Enteroaggregative *Escherichia coli. Emerg Infect Dis*, 4(2), 251–261.

Neves, D., Job, V., Dortet, L., Cossart, P., Dessen, A. (2013). Structure of internalin InlK from the human pathogen Listeria monocytogenes. *J Mol Biol*, 425(22), 4520–4529.

Newell, D.G., Koopmans, M., Verhoef, L., Duizer, E., Aidara-Kane, A., Sprong, H., Opsteegh, M., Langelaar, M., Threfall, J., Scheutz, F., Van Der Giessen, J., Kruse, H. (2010). Food-borne diseases – The challenges of 20 years ago still persist while new ones continue to emerge. *Int J Food Microbiol*, 139(Suppl 1), S3–S15.

Niess, J.H., Brand, S., Gu, X., Landsman, L., Jung, S., McCormick, B.A., Vyas, J.M., Boes, M., Ploegh, H.L., Fox, J.G., Littman, D.R., Reinecker, H.-C. (2005). CX_3CR1-mediated dendritic cell access to the intestinal lumen and bacterial clearance. *Science*, 307(5707), 254–258.

Nikitas, G., Deschamps, C., Disson, O., Niault, T., Cossart, P., Lecuit, M. (2011). Transcytosis of *Listeria monocytogenes* across the intestinal barrier upon specific targeting of goblet cell accessible E-cadherin. *J Exp Med*, 208(11), 2263–2277.

Nougayrede, J.P., Homburg, S., Taieb, F., Boury, M., Brzuszkiewicz, E., Gottschalk, G., Buchrieser, C., Hacker, J., Dobrindt, U., Oswald, E. (2006). *Escherichia coli* induces DNA double-strand breaks in eukaryotic cells. *Science*, 313(5788), 848–851.

Novik, V., Hofreuter, D., Galán, J.E. (2010). Identification of *Campylobacter jejuni* genes involved in its interaction with epithelial cells. *Infect Immun*, 78(8), 3540–3553.

Ó Cróinín, T. and Backert, S. (2012). Host epithelial cell invasion by *Campylobacter jejuni*: Trigger or zipper mechanism? *Front Cell Infect Microbiol*, 2.

Odeyemi, O.A. (2016). Public health implications of microbial food safety and foodborne diseases in developing countries. *Food Nutr Res*, 60, 29819.

Peck, M.W., Smith, T.J., Anniballi, F., Austin, J.W., Bano, L., Bradshaw, M., Cuervo, P., Cheng, L.W., Derman, Y., Dorner, B.G., Fisher, A., Hill, K.K., Kalb, S.R., Korkeala, H., Lindström, M., Lista, F., Lúquez, C., Mazuet, C., Pirazzini, M., Popoff, M.R., Rossetto, O., Rummel, A., Sesardic, D., Singh, B.R., Stringer, S.C. (2017). Historical perspectives and guidelines for botulinum neurotoxin subtype nomenclature. *Toxins*, 9(1), 38.

Penheiter, K.L., Mathur, N., Giles, D., Fahlen, T., Jones, B.D. (1997). Non-invasive *Salmonella* Typhimurium mutants are avirulent because of an inability to enter and destroy M cells of ileal Peyer's patches. *Mol Microbiol*, 24(4), 697–709.

Pentecost, M., Otto, G., Theriot, J.A., Amieva, M.R. (2006). *Listeria monocytogenes* invades the epithelial junctions at sites of cell extrusion. *PLoS Pathog*, 2(1), e3.

Pirazzini, M., Rossetto, O., Eleopra, R., Montecucco, C. (2017). Botulinum neurotoxins: Biology, pharmacology, and toxicology. *Pharmaco Rev*, 69(2), 200–235.

Pizarro-Cerdá, J. and Cossart, P. (2009). *Listeria monocytogenes* membrane trafficking and lifestyle: The exception or the rule? *Annu Rev Cell Deve Biol*, 25(1), 649–670.

Ponniah, J., Robin, T., Margaret, S., Son, R., Farinazleen, M.G., Cheah, Y.K. (2010). Detection of *Listeria monocytogenes* in foods. *Int Food Res J*, 17, 1–11.

Pradhan, S., Karve, S.S., Weiss, A.A., Hawkins, J., Poling, H.M., Helmrath, M.A., Wells, J.M., McCauley, H.A. (2020). Tissue responses to Shiga toxin in human intestinal organoids. *Cell Mole Gastroenterol Hepatol*, 10(1), 171–190.

Purdy, D., Buswell, C.M., Hodgson, A.E., McAlpine, K., Henderson, I., Leach, S.A. (2000). Characterisation of cytolethal distending toxin (CDT) mutants of *Campylobacter jejuni*. *J Med Microbiol*, 49(5), 473–479.

Raffatellu, M., George, M.D., Akiyama, Y., Hornsby, M.J., Nuccio, S.-P., Paixao, T.A., Butler, B.P., Chu, H., Santos, R.L., Berger, T., Mak, T.W., Tsolis, R.M., Bevins, C.L., Solnick, J.V., Dandekar, S., Bäumler, A.J. (2009). Lipocalin-2 resistance confers an advantage to *Salmonella enterica* Serotype Typhimurium for growth and survival in the inflamed intestine. *Cell Host Microbe*, 5(5), 476–486.

Rajkowski, K.T. and Smith, J.L. (2001). Update: Food poisoning and other diseases induced by *Bacillus cereus*. In *Foodborne Disease Handbook*, Hui, Y.H., Pierson, M.D., Gorham, J.R. (eds). Markel Dekker, New York.

Rescigno, M., Urbano, M., Valzasina, B., Francolini, M., Rotta, G., Bonasio, R., Granucci, F., Kraehenbuhl, J.P., Ricciardi-Castagnoli, P. (2001). Dendritic cells express tight junction proteins and penetrate gut epithelial monolayers to sample bacteria. *Nat Immunol*, 2(4), 361–367.

Richardson, S.E., Karmali, M.A., Becker, L.E., Smith, C.R. (1988). The histopathology of the hemolytic uremic syndrome associated with verocytotoxin-producing *Escherichia coli* infections. *Hum Pathol*, 19(9), 1102–1108.

Rivera-Amill, V., Kim, B.J., Seshu, J., Konkel, M.E. (2001). Secretion of the virulence-associated *Campylobacter* invasion antigens from *Campylobacter jejuni* requires a stimulatory signal. *J Infect Dis*, 183(11), 1607–1616.

Robinson, D.A. (1981). Infective dose of *Campylobacter jejuni* in milk. *B Med J*, 282(6276), 1584–1584.

Rosselin, M., Virlogeux-Payant, I., Roy, C., Bottreau, E., Sizaret, P.Y., Mijouin, L., Germon, P., Caron, E., Velge, P., Wiedemann, A. (2010). Rck of *Salmonella enterica*, subspecies enterica serovar enteritidis, mediates zipper-like internalization. *Cell Res*, 20(6), 647–664.

Ruggenenti, P. and Remuzzi, G. (1998). Pathophysiology and management of thrombotic microangiopathies. *J Nephrol*, 11(6), 300–310.

Salimnia, H., Patel, D., Lephart, P.R., Fairfax, M.R., Chandrasekar, P.H. (2010). *Listeria grayi*: Vancomycin-resistant, gram-positive rod causing bacteremia in a stem cell transplant recipient. *Transpl Infect Dis*, 12(6), 526–528.

Scallan, E., Griffin, P.M., Angulo, F.J., Tauxe, R.V., Hoekstra, R.M. (2011a). Foodborne illness acquired in the United States unspecified agents. *Emerg Infect Dis J*, 17(1), 16.

Scallan, E., Hoekstra, R.M., Angulo, F.J., Tauxe, R.V., Widdowson, M.-A., Roy, S.L., Jones, J.L., Griffin, P.M. (2011b). Foodborne illness acquired in the United States – Major pathogens. *Emerg Infect Dis*, 17(1), 7–15.

Schlumberger, M.C. and Hardt, W.-D. (2006). *Salmonella* type III secretion effectors: Pulling the host cell's strings. *Curr Opin Microbiol*, 9(1), 46–54.

Schüller, S., Frankel, G., Phillips, A.D. (2004). Interaction of Shiga toxin from *Escherichia coli* with human intestinal epithelial cell lines and explants: Stx2 induces epithelial damage in organ culture. *Cell Microbiol*, 6(3), 289–301.

Shen, Y., Naujokas, M., Park, M., Ireton, K. (2000). InlB-dependent internalization of *Listeria* is mediated by the Met receptor tyrosine kinase. *Cell*, 103(3), 501–510.

Silva, J., Leite, D., Fernandes, M., Mena, C., Gibbs, P.A., Teixeira, P. (2011). *Campylobacter* spp. as a foodborne pathogen: A review. *Front Microbiol*, 2, 200.

Stenfors Arnesen, L.P., Fagerlund, A., Granum, P.E. (2008). From soil to gut: *Bacillus cereus* and its food poisoning toxins. *FEMS Microbiol Rev*, 32(4), 579–606.

Su, L.H. and Chiu, C.H. (2007). *Salmonella*: Clinical importance and evolution of nomenclature. *Chang Gung Med J*, 30(3), 210–219.

Sun, H., Chow, E.C.Y., Liu, S., Du, Y., Pang, K.S. (2008). The Caco-2 cell monolayer: Usefulness and limitations. *Expert Opin Drug Metab Toxicol*, 4(4), 395–411.

Tarr, P.I. (2009). Shiga toxin-associated hemolytic uremic syndrome and thrombotic thrombocytopenic purpura: Distinct mechanisms of pathogenesis. *Kidney Int*, 75, S29–S32.

Tarr, P.I., Gordon, C.A., Chandler, W.L. (2005). Shiga-toxin-producing *Escherichia coli* and haemolytic uraemic syndrome. *Lancet*, 365(9464), 1073–1086.

Taylor, D.N., Bopp, C., Birkness, K., Cohen, M.L. (1984). An outbreak of salmonellosis associated with a fatality in a healthy child: A large dose and severe illness. *Am J Epidemiol*, 119(6), 907–912.

Tegtmeyer, N., Rohde, M., Backert, S. (2012). Clinical presentations and pathogenicity mechanisms of bacterial foodborne infections. In *Microbial Food Safety*, Oyarzabal, O.A. and Backert, S. (eds). Springer, New York.

Threlfall, E.J. (2002). Antimicrobial drug resistance in *Salmonella*: Problems and perspectives in food- and water-borne infections. *FEMS Microbiol Rev*, 26(2), 141–148.

Tong, S.Y.C., Davis, J.S., Eichenberger, E., Holland, T.L., Fowler, V.G. (2015). *Staphylococcus aureus* infections: Epidemiology, pathophysiology, clinical manifestations, and management. *Clin Microbiol Rev*, 28(3), 603–661.

Uzzau, S., Brown, D.J., Wallis, T., Rubino, S., Leori, G., Bernard, S., Casadesús, J., Platt, D.J., Olsen, J.E. (2000). Host adapted serotypes of *Salmonella enterica*. *Epidemiol Infect*, 125(2), 229–255.

Vázquez-Boland, J.A., Kuhn, M., Berche, P., Chakraborty, T., Domínguez-Bernal, G., Goebel, W., González-Zorn, B., Wehland, J., Kreft, J. (2001). *Listeria* pathogenesis and molecular virulence determinants. *Clin Microbiol Rev*, 14(3), 584–640.

Vazquez-Torres, A., Jones-Carson, J., Bäumler, A.J., Falkow, S., Valdivia, R., Brown, W., Le, M., Berggren, R., Parks, W.T., Fang, F.C. (1999). Extraintestinal dissemination of *Salmonella* by CD18-expressing phagocytes. *Nature*, 401(6755), 804–808.

Vega, F.E., Kaya, H.K., Tanada, Y. (2012). *Insect Pathology*, 2nd edition. Elsevier and Academic Press, Amsterdam.

Velge, P., Wiedemann, A., Rosselin, M., Abed, N., Boumart, Z., Chaussé, A.M., Grépinet, O., Namdari, F., Roche, S.M., Rossignol, A., Virlogeux-Payant, I. (2012). Multiplicity of *Salmonella* entry mechanisms, a new paradigm for *Salmonella* pathogenesis. *MicrobiologyOpen*, 1(3), 243–258.

Vernay, T., Cannie, I., Gaboriau, F., Gall, S.D., Tamanai-Shacoori, Z., Burel, A., Jolivet-Gougeon, A., Loréal, O., Bousarghin, L. (2020). Bacteroides fragilis prevents *Salmonella* Heidelberg translocation in co-culture model mimicking intestinal epithelium. *Benef Microbes*, 11(4), 391–401.

Vernozy-Rozand, C. (2005). *Escherichia coli* producteurs de shigatoxines (STEC). In *Bactériologie alimentaire, compendium d'hygiène des aliments*, Federighi, M. (ed.). Economica, Paris.

Von Martels, J.Z.H., Sadaghian Sadabad, M., Bourgonje, A.R., Blokzijl, T., Dijkstra, G., Faber, K.N., Harmsen, H.J.M. (2017). The role of gut microbiota in health and disease: *In vitro* modeling of host–microbe interactions at the aerobe–anaerobe interphase of the human gut. *Anaerobe*, 44, 3–12.

Watson, R.O. and Galán, J.E. (2008). *Campylobacter jejuni* survives within epithelial cells by avoiding delivery to lysosomes. *PLoS Pathog*, 4(1), e14.

WHO (2015). WHO estimates of the global burden of foodborne diseases: Foodborne diseases burden epidemiology reference group 2007–2015. Report, WHO.

WHO (2017). The burden of foodborne diseases in the WHO European region. Report, WHO.

Yamasaki, S., Asakura, M., Tsukamoto, T., Faruque, S.M., Deb, R., Ramamurthy, T. (2006). Cytholethal distending toxin (CDT): Genetic diversity, structure and role in diarrheal disease. *Toxin Rev*, 25(1), 61–88.

Yao, R., Burr, D.H., Guerry, P. (1997). CheY-mediated modulation of *Campylobacter jejuni* virulence. *Mol Microbiol*, 23(5), 1021–1031.

Zamora, C.Y., Ward, E.M., Kester, J.C., Kelly Chen, W.L., Velazquez, J.G., Griffith, L.G., Imperiali, B. (2019). Application of a gut-immune co-culture system for the study of N-glycan-dependent host–pathogen interactions of *Campylobacter jejuni*. *Glycobiol*, 30, 374–81.

5

Microbial Dose–Response Models

Moez SANAA and Laurent GUILLIER
ANSES, Department of Risk Assessment, Maison-Alfort, France

5.1. Introduction

Risk assessment is a key component in ensuring that scientific criteria are used to establish food safety standards, guidelines and other recommendations to ensure consumer protection and facilitate international trade. Risk assessment is a structured and transparent process for organizing available data. It is designed to model the transmission of the pathogen by considering the different stages of the food chain: "from farm to fork". Quantitative microbial risk assessment consists of several steps. Exposure assessment estimates the distribution of exposure in a human population to a specific pathogenic microorganism through one or more foods. Dose–response (DR) modeling is undertaken to quantify the likelihood of infection and disease as a function of ingested dose. Risk characterization uses the results of exposure assessment and DR modeling to derive an estimate of the risk of disease in the population, expressed by year, by consumption occasion or otherwise.

Ingestion of pathogenic bacteria does not always result in infection, illness or death. Various factors can act as barriers or increase the likelihood of infection or disease. In assessing the effects, several factors related to the microorganism, the human host and the environment should be considered. The ability to multiply, the infectious capacity and the virulence of a

Microbiological Risk Assessment Associated with the Food Processing and Distribution Chain,
coordinated by Jeanne-Marie MEMBRÉ. © ISTE Ltd 2022.

microorganism are the result of its interaction with the host and the environment (food matrix). For example, genetic material can be transferred from one microorganism to another, resulting in the transfer of certain characteristics such as antibiotic resistance and virulence factors; food characteristics, such as fat content, can alter pathogenicity. The food manufacturing process can cause heat shock, sometimes leading to adaptation of the microorganism, such as its acid tolerance, resulting in an increased probability of the bacteria surviving in the stomach of the consumer. Regarding the host, we can cite as an example the increased sensitivity depending on age, health status, ingestion of certain drugs, immune status, etc.

Infection can be seen as the process of overcoming various host barriers, which may or may not be equally effective in eliminating or inactivating pathogenic microorganisms. Each individual microorganism has a certain likelihood or relative frequency of crossing a barrier.

The infection process can be broken down into two stages:

– in the first, the host ingests bacteria capable of causing infection or disease;

– in the second, a fraction of the ingested bacteria arrive at a site where infection can begin.

The probability of infection can be written as:

$$Probability\ (n|d) \times Probability\ (k|n)$$

where n is the number of microorganisms actually being ingested, knowing that the food has an average concentration of microorganisms equal to d, and k is the number of microorganisms that pass through all the barriers and cause infection.

In this chapter, we present the main mathematical models that can be used to describe this infection process and their respective parameters, discuss the main approaches for estimating the parameters of the main models based on experimental or epidemiological data and finally demonstrate the ways these models can be used in quantitative risk assessment or in the setting of health-based guide values.

5.2. Main dose–response models

Dose–response modeling is the process of using mathematical relationships to describe the probability of an adverse health effect (e.g. infection or illness) occurring in an individual or the frequency of an adverse health effect in a population when that individual or population is exposed to a specific dose of pathogenic microorganisms. The dose level may be measured in terms of a discrete number of organisms (e.g. oocysts), colony-forming units (CFUs) or plaque-forming units (PFUs) or by molecular methods.

Regardless of how microorganisms are measured, we consider in this presentation that an individual named "i" is being exposed to an exact number of units x of a pathogenic microorganism. The probability of observing an infection or disease following this exposure will depend on the number of microorganism units k among the x ingested microorganism units surviving the defensive barriers of the individual "i".

At this stage of the process, we can already distinguish two types of models (Haas 2002):

– models that assume a cooperation between ingested microorganisms, involving the notion of minimum infective dose: the infection is the result of the combined action of multiple microorganism units, and it becomes possible only if k exceeds a threshold k_{min};

– threshold-free models, which assume independence in the action of microorganism units. Each microorganism unit has a non-zero chance of causing infection on its own.

If we assume that the x ingested units have the same probability of survival, that is, of crossing the gastrointestinal and immunological barriers (*p*), then the probability distribution of the number of surviving units (infection-initiating) is binomial. If only one bacterium can cause infection, the dose–response relationship is equal to the complement of the probability that no units survive:

$$P(\text{infection}[x) = 1 - (1 - p)^x$$

The model becomes more complicated if the minimum dose assumption ($k_{\min}$) is assumed:

$$P(\text{infection}[x) = 1 - \sum_{j=0}^{k_{min}-1} C_x^j p^j (1-p)^{x-j}$$

The main models in microbiology assume the absence of a minimum dose.

For some parasites, like *Trichinella*, the infection process requires the survival of at least two larvae of different sexes. The dose–response model adopted for *Trichinella* assumes the independence of action of ingested larvae while taking into account the parasite's need to reproduce in the host's intestine. Occurrence of infection or disease first requires the survival of at least one male–female pair. The female lays larvae, which are capable of crossing the intestine and migrating to the parasite's predilection sites. The probability of survival is considered to be the same for male and female larvae (p). If the proportion of females is r, then the probability of infection after the ingestion of N larvae is equal to the probability that at least one female larva and one male larva will survive (Teunis *et al.* 2012):

$$P(infection) = P(k_F > 0) \times P(k_M > 0)$$

where k_F and k_M are the number of surviving females and number of surviving males capable of reproducing, respectively. They both have binomial distributions with numbers of trials N_F and N_M, respectively, and a probability of success equal to p. N_F and N_M are the numbers of females and males in the N ingested larvae, respectively. The N_F number has a binomial distribution with parameters N ($N = N_F + N_M$) and r.

The probability p_m of survival for a larva is the result of interactions between the specific factors of the host, parasite and food. It is considered the same for every exposure occasion, that is, for the group of larvae ingested when a given individual consumes a food. However, this probability will vary between various exposure occasions and can therefore have values between 0 and 1. This variability is described by a beta distribution with the strictly positive parameters α and β. The available epidemiological data are used to estimate these last two parameters. The probability of infection as a function of the ingestion of N_F female larvae and N_M male larvae can be deduced based on the parameters of the beta distribution.

5.2.1. *Exponential model*

The exponential model is a no-threshold model, where the probability of bacterial survival and initiation of infection (p) is constant, and the number of bacteria n ingested is not perfectly known but assumed to follow a Poisson distribution of parameter μ (Figure 5.1).

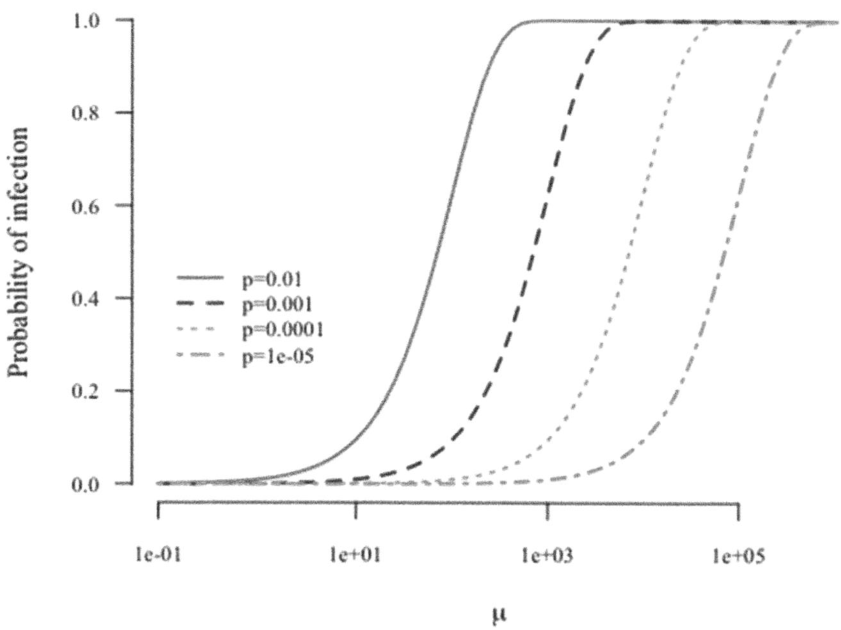

Figure 5.1. *Examples of exponential distribution. For a color version of this figure, see www.iste.co.uk/membre/microbiological.zip*

The hypothesis of the constancy of p means that the result of the interaction between host, microorganism and environmental factors does not vary much between exposed individuals. This is to be expected in experimental situations where, for example, animals of the same lineage, age and sex consume the same food contaminated with the same strain of the pathogenic microorganism, or in food poisoning situations affecting a homogeneous community where the exposure comes from the same food and the same strain of the microorganism.

The Poisson distribution is applicable in cases in which the contamination of the food is considered homogeneous. This assumption is perfectly suitable for liquid foods and when the units of the microorganisms do not aggregate together to produce clusters.

The probability of being infected following the ingestion of a food serving in which the average contamination is equal to μ is expressed as follows:

$$P(\textit{infection}|\mu) = 1 - exp(-p \times \mu)$$

5.2.2. *Beta-Poisson model*

The hypothesis of the unchangeable parameter p of the exponential model makes it difficult to extrapolate this model to the general population. In fact, in the general population, individuals react differently to exposure to a certain pathogenic strain depending on their age, their health status and the development of their immune system. Moreover, strains of the same pathogenic organism may have very different characteristics that may increase or decrease their ability to invade the host organism. These differences may be the result of the genetic selection during the transmission of pathogenic strains from one person to another or because of the conditions they encounter along the food chain. These two sources of variability can be combined and described using a single probability distribution. The beta distribution is particularly suitable for describing the variability of the parameter p. Indeed, the beta distribution is a continuous distribution defined in the interval [0, 1] with two strictly positive parameters, denoted α and β, which allow the distribution to take very diversified shapes (Figure 5.2).

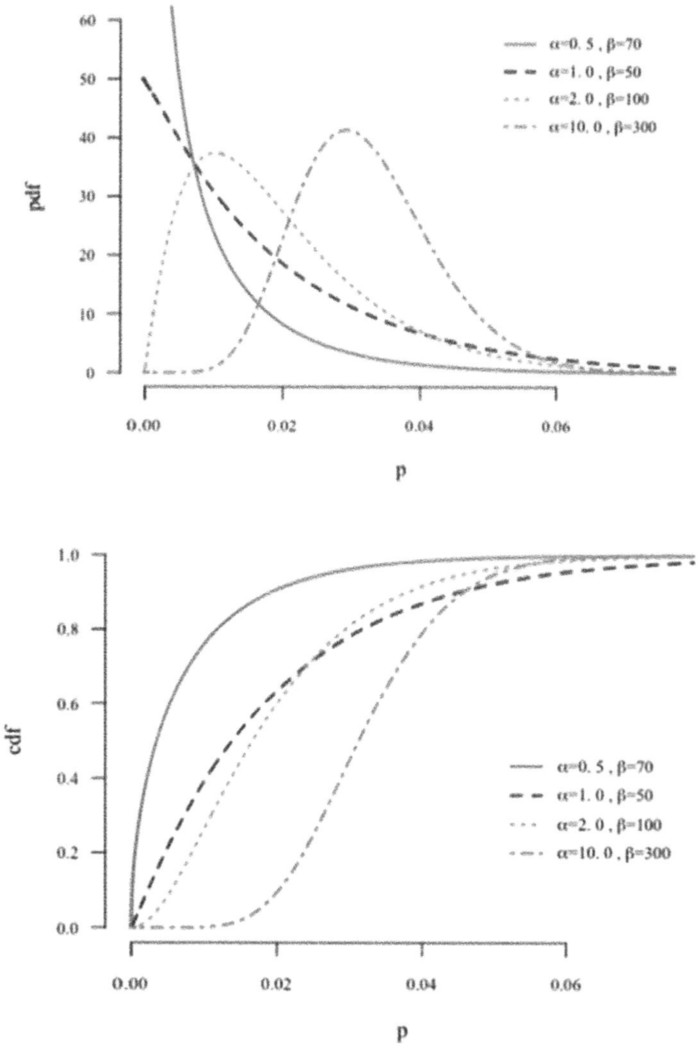

Figure 5.2. *Examples of beta distribution. Top: probability density function (pdf); bottom: cumulative distribution function (cdf). For a color version of this figure, see www.iste.co.uk/membre/microbiological.zip*

With the assumption that p is variable and well-described by a beta distribution, we shift to a two-component hierarchical model: a first random component, which consists of randomly drawing a value p from the beta distribution, and a second component, which is the same as using the exponential model dependent on p:

$$\begin{cases} p_i \sim Beta(\alpha, \beta) \\ P(\text{infection}|\mu, p_i) = 1 - \exp(-p_i \times \mu) \end{cases}$$

This hierarchical model therefore allows us to assign to each individual of the population a specific dose–response relationship. In practice, however, it is rather rare that the objective of a risk assessment is to estimate the risk for a particular individual. Most often we seek to estimate the probability for the population as a whole; in other words, it is to estimate the average probability of infection or disease for the whole population as a function of the dose:

$$\begin{aligned} P(\mathrm{I}|\mu, \alpha, \beta) &= 1 - \int_{p=0}^{p=1} \exp(-p \times \mu) \times Beta(p; \alpha, \beta)\, dp \\ &= 1 - 1F1(\alpha, \beta, -d_i) \end{aligned}$$

where $Beta(p; \alpha, \beta)$ is the probability density function of the beta distribution with the parameters α and β, and $1F1$ is the confluent hypergeometric function.

The confluent hypergeometric function is very hard to approximate. Different calculation approaches have been implemented for the use of this function, for example, the BAS package in R (Clyde 2018). However, there is an approximation of this function acceptable under certain conditions for the parameter values of the distribution, $\alpha << \beta$ and $\beta >> 1$:

$$P(d_i) = 1 - \left(1 + \frac{\mu}{\beta}\right)^{-\alpha}$$

Before using this approximation of the confluent function, the following rule of thumb should be used (Xie *et al.* 2017):

$$\beta > \sqrt{22\alpha} \text{ or } 0.02 < \alpha < 2$$

Figures 5.3 and 5.4 show examples of beta-Poisson dose–response models and how the parameters influence the shape of the function.

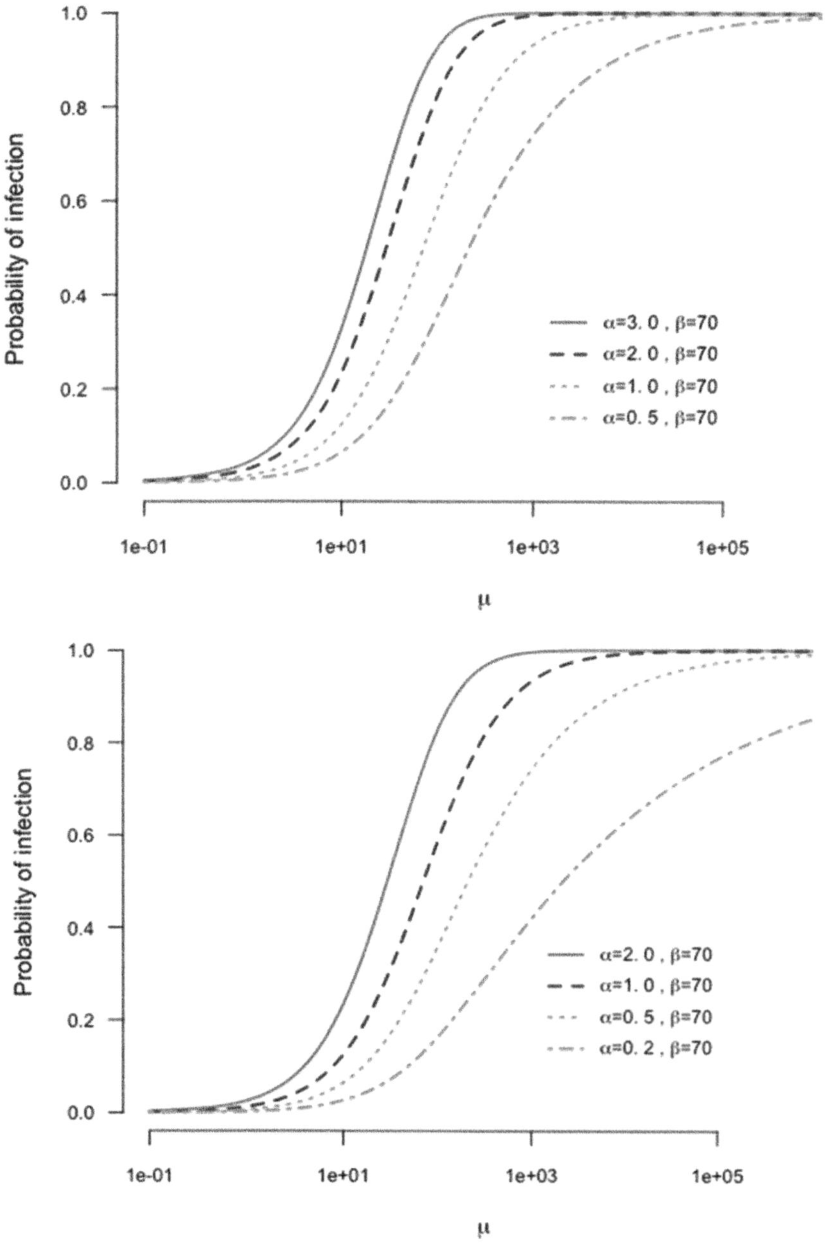

Figure 5.3. *Examples of beta-Poisson dose–response models. For a color version of this figure, see www.iste.co.uk/membre/microbiological.zip*

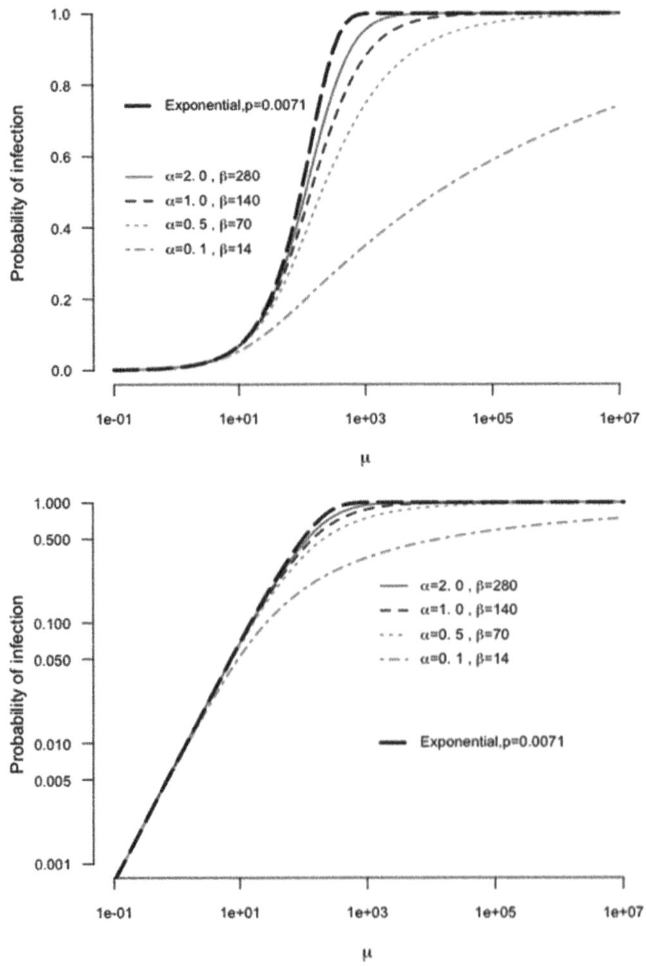

Figure 5.4. *Examples of beta-Poisson dose–response models. The parameters α and β were chosen to obtain the same arithmetic mean for the beta distributions ($p = \alpha/(\alpha+\beta) = 0.0071$). Top: arithmetic scale; bottom: logarithmic scale. When α is higher than 1, the exponential model using the arithmetic mean of the beta distribution can be used as an approximation of the beta-Poisson model. For a color version of this figure, see www.iste.co.uk/membre/microbiological.zip*

5.2.3. *Beta-negative binomial model*

In some situations, individuals are exposed to pathogens via solid food under contamination conditions that do not always lead to a homogeneous distribution of the microorganisms. The units of the pathogen may be unequally distributed in the food portions and therefore the Poisson distribution is no longer appropriate. A good alternative to the Poisson distribution is the negative binomial distribution. The probability of being actually exposed to x units of the pathogen is obtained using the following function:

$$f(x) = \frac{\Gamma(x+s)}{\Gamma(s)x!}\left(\frac{s}{\mu+s}\right)^{s}\left(\frac{\mu}{\mu+s}\right)^{x}$$

where μ and s are the mean and the shape parameter of the negative binomial distribution, respectively, and $\Gamma(.)$ is the gamma function. The variance of the negative binomial distribution is:

$$Var(X) = \mu + \frac{\mu^2}{s}$$

The shape parameter s controls the homogeneity of the contamination. When s tends to infinity, the variance converges to the mean of the distribution and thus the distribution can be approximated by a Poisson distribution.

If we assume that the pathogen follows a negative binomial distribution with the parameters μ and s, then the expected number of microorganisms per unit will follow a Gamma distribution with the same parameters μ and s (Zelterman 2005). The overall probability of illness assuming an exponential dose–response model is then obtained by:

$$Prob(I|\mu, s) = 1 - \int_0^{\infty} \exp(-px)\, g(x)dx$$

where g(x) is the probability density function of the Gamma distribution. The integral is equivalent to the moment-generating function of the Gamma distribution, $E[exp(tx)]$ (Zelterman 2005), which has an analytical solution. The dose–response when assuming a negative binomial distribution of the pathogen in food can be expressed as:

$$Prob(I|\mu, s) = 1 - \left(1 + \frac{p\mu}{s}\right)^{-s}$$

In the case where the exponential model is not appropriate and where the parameter p of the dose–response relationship among individuals is sought as variable, typically according to a beta distribution with parameters (α, β), the dose–response model evolves to the beta-negative binomial dose–response model:

$$Prob(I|\mu, s, \alpha, \beta) = 1 - \int_0^1 \left(1 + \frac{p\mu}{s}\right)^{-s} f(p; \alpha, \beta)\, dp$$

where $f(p, \alpha, \beta)$ is the probability density function of the beta distribution with parameters (α, β).

5.2.4. *Models based on the log-normal distribution*

Instead of the beta distribution, the log-normal distribution can be used to describe the variability of the parameter p, the probability of survival of a unit of the pathogen against the different defense barriers of the host organism. The advantage of this choice lies in the possibility of easily decomposing the variance at the log scale. Indeed, the variability of p can be associated on the one hand with the diversity of the strains of the pathogen – different levels of virulence – and on the other hand with the diversity of the host organisms – different capacities to face the invasion of a pathogen. This has been used to generate a dose–response model for *Listeria monocytogenes* (Pouillot *et al.* 2015) in which log (p)~$Normal(\mu,\sigma)$. Epidemiological data on *L. monocytogenes* show a large variability in susceptibility in the population. One of the main risk factors for the occurrence of listeriosis is the presence of preconditions (Leclercq *et al.* 2020) such as pregnancy, non-hematological or hematological cancer, renal or liver failure (dialysis, cirrhosis), a solid organ transplant, inflammatory diseases (rheumatoid arthritis, ulcerative colitis, giant cell arteritis, Crohn's disease), HIV/AIDS, diabetes (type I or type II) or heart diseases. Given the wide range of increased risk between individuals with any of these conditions and the population without preconditions and under 65 years of age, relative risks ranging from 5.5 to 373 (Goulet *et al.* 2012), it is convenient to stratify the population according to these different types of pre-conditions (subpopulations). The model assumes that the variability between the mean (μ) is specific to each of the subpopulations considered and the standard deviation (σ) is assumed to be the same for all population segments. It includes the variability among *L. monocytogenes* strains (σ^2_{sv}) and individual host sensitivities (σ^2_{host}), $\sigma^2 = \sigma^2_{sv} + \sigma^2_{host}$. Twelve percent of the overall variability of p is attributed to strain variability, with the remainder being the variability of host individuals within each subpopulation. Standard deviations were estimated using animal tests and data from the FDA/FSIS risk assessment. Figure 5.5 shows the dose–response models for each of the 11 sub-populations considered (Pouillot *et al.* 2015).

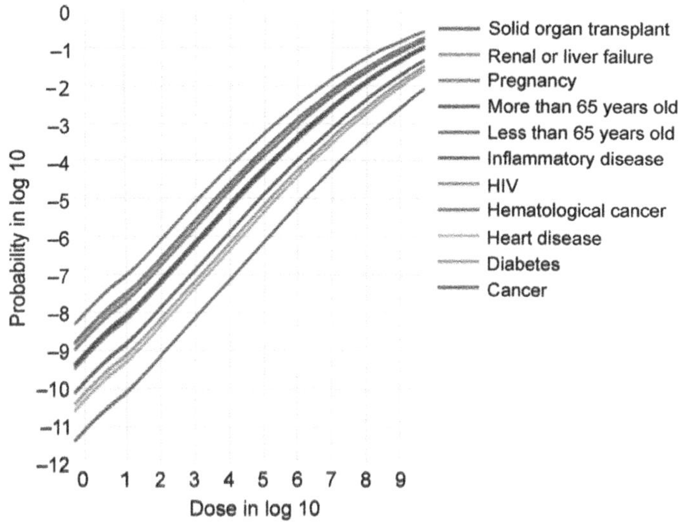

Figure 5.5. *Dose–response models (probability of severe listeriosis cases conditional on the exposed dose) for each of the 11 population segments considered by Pouillot* et al. *(2015). For a color version of this figure, see www.iste.co.uk/membre/microbiological.zip*

5.3. Fitting data to a dose–response model

5.3.1. *Experimental studies*

The most common practice for dose–response modeling is to fit limited datasets that have been derived from experimental feeding studies to statistical models that are often biologically based.

The experimental design of a feeding study enrolls in general volunteers who are assigned to different groups receiving increasing doses of the pathogen. As an example, we use the data first reported by Black *et al.* (1988) and analyzed by Medema *et al.* (1996). In this study, adult volunteers ingested different doses of a strain of *Campylobacter jejuni* (A3249). The inoculum was prepared in non-fat milk, and the doses ranged from 810 to 10^8 (Table 5.1).

Dose levels	Infected	Non-infected	Total
810	5	5	10
8100	6	4	10
91000	11	2	13
810000	8	3	11
1.1E+06	15	4	19
1.1E+08	5	0	5

Table 5.1. *Dose levels and response to* Campylobacter jejuni *(A3249) (Black* et al. *1988)*

The fitting of the data to a dose response is based on an initial hypothesis that the number of responses, in this case "infection", is generated from a binomial distribution:

$$y_i \sim Binomial(\pi_i, n_i)$$

where π_i is the probability of infection as a function of the dose, f(dose). In this type of design, the dose is not known for each individual, but we assume given the preparation of the inoculum that it follows a Poisson distribution for the μ_i parameter (dose levels in the table). π_i is specific to each of the group of n_i individuals exposed to a dose level μ_i and can be linked to the dose using the exponential or beta-Poisson model:

$$\pi_i = f(\mu_i, \theta)$$

where θ represents the parameters of the dose–response model and i from 1 to m describes the tested dose levels (in Table 5.1, m = 6). The method generally used to estimate model parameters is maximum likelihood estimation. The likelihood of the model, the probability of observing the responses, is written as follows:

$$L_m = \prod_{i=1}^{m} [f(\mu_i, \theta)]^{y_i} [1 - f(\mu_i, \theta)]^{n_i - y_i}$$

The method's principle is to study the variation of the likelihood as a function of the values of the parameters θ and to take as an estimator those that maximize the likelihood. Analytically, after logarithmic transformation, we look for the solution of the system of partial derivatives with respect to the parameters of the model. However, the equation system has no analytical solution for optimizing the likelihood. Numerical methods, such as the Newton–Raphson method implemented in R (*mle( )* function) (Mazucheli *et al.* 2017), can be used to estimate the dose–response model parameters.

Parameters	Estimates	Standard errors	Deviance*
Exponential			110.11
$\ln(\hat{p})$	–12.617	0.177	
Beta-Poisson			2.43
$\ln(\hat{\alpha})$	–1.936	0.435	
$\ln(\hat{\beta})$	1.995	3.681	

*Deviance is minus two times the logarithm of the likelihood function

Table 5.2. *Estimates and standard errors for the parameters from both the exponential and beta-Poisson models using the feeding study data presented in Table 5.1*

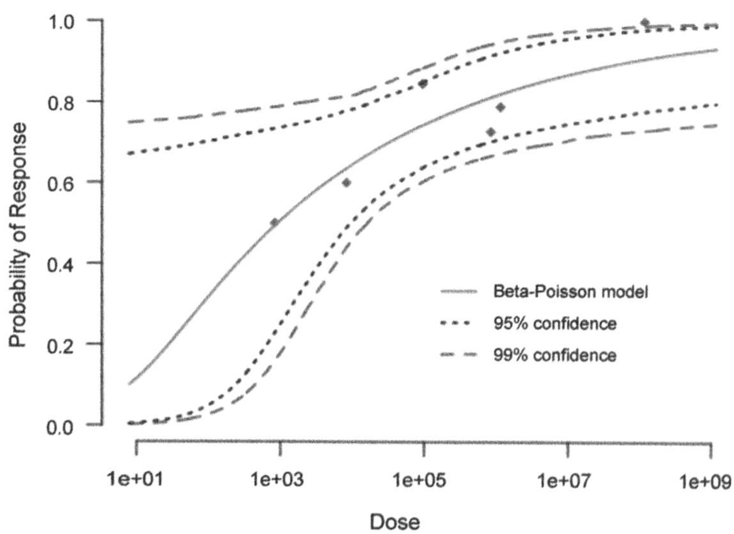

Figure 5.6. *Best-fitting curve from the beta-Poisson model and the 95% and 99% confidence range from 10,000 bootstrap iterations. For a color version of this figure, see www.iste.co.uk/membre/microbiological.zip*

The beta-Poisson model fits the data better than the Poisson model, because its deviance is less than the deviance obtained using the Poisson model (Table 5.2). The maximum likelihood approach provides the parameters' estimates and their standard deviations. Given the possible correlation between the parameters of the model as well as the small number of observations, it is not possible to calculate directly the confidence intervals of the parameters' estimates. By re-sampling the observed data, it is possible to generate several possible sets of observations (bootstrapping) and to re-apply the maximum-likelihood-estimation algorithm several times (10,000). We can obtain a collection of dose–response curves, and then it is possible to compute the needed confidence intervals (Figures 5.6 and 5.7).

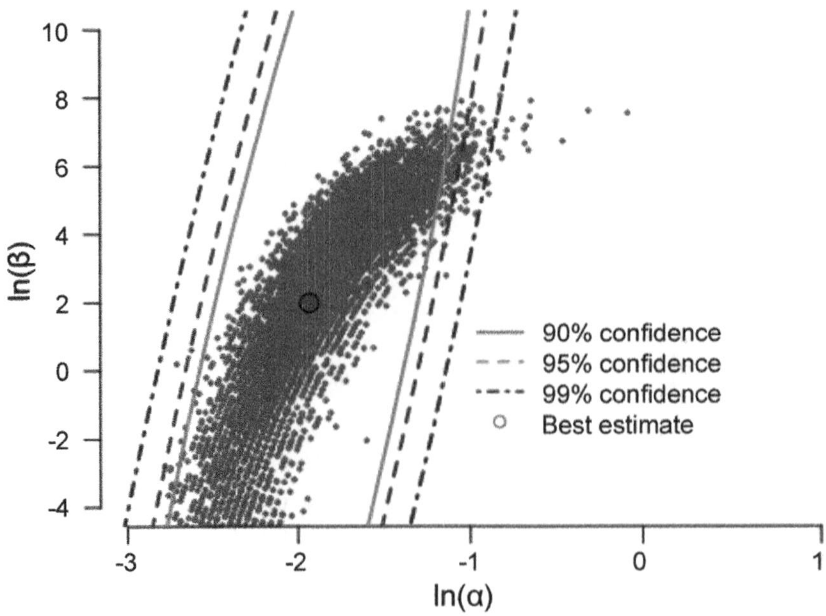

Figure 5.7. *Beta-Poisson bootstrap estimates (10,000 iterations – data in Table 5.1). For a color version of this figure, see www.iste.co.uk/membre/microbiological.zip*

5.3.2. *Outbreak data*

Data from foodborne outbreak investigations can be used to develop dose–response models. To illustrate the approach, let us consider a large community outbreak of hemolytic uremic syndrome (HUS) caused by *E. coli* 0157 in France and associated with the consumption of contaminated frozen ground beef patties in households (Delignette-Muller *et al.* 2008). The idea is to first leverage epidemiological estimates of HUS incidence rates for different age groups in the population (King *et al.* 2009) and data collected at the time of the study. Fifteen age groups were defined by King *et al.*: 1 to 15 years. Children under six months of age were excluded because they were expected to have a very different dietary diversification than the other age groups. From King *et al.* (2009), the 16 incidence rates (IRs) per 100,000 children were 1.82, 2.88, 2.2, 1.46, 0.91, 0.62, 0.46, 0.36, 0.22, 0.16, 0.19, 0.13, 0.09, 0.08, 0.05 and 0.04. Combining the observed incidence rates and subpopulation sizes (excluding those of the children less than six months old because they were expected to have a very different diet compared to other ages), the trend in HUS incidence as a function of age (Figure 5.8) is:

$$\text{Log}(\text{IR} \mid \text{age}) = \text{IR}_0 + \text{k} \times \text{age}$$

$\text{IR}_0 = -2.21$ with 95% credible interval [–2.31, –2.11] and $k = -0.38$ with 95% credible interval [–0.40, –0.36]

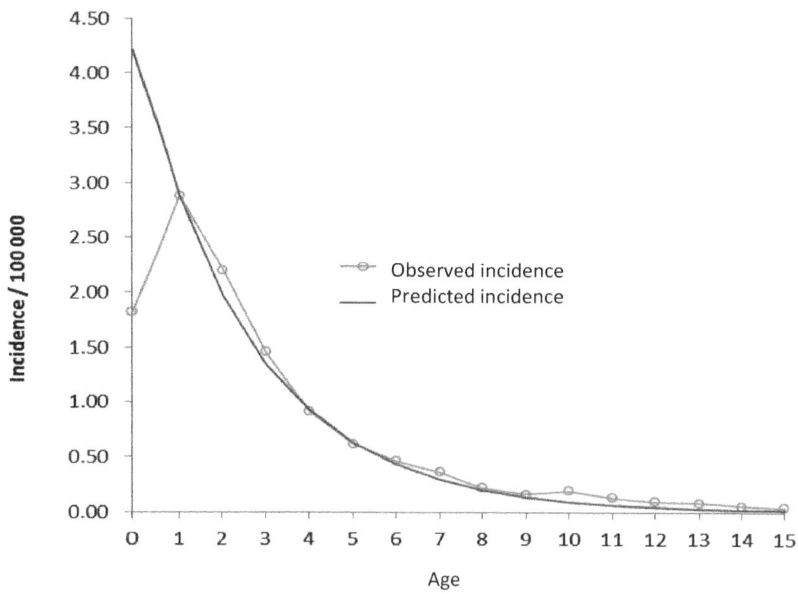

Figure 5.8. *Incidence rates of hemolytic uremic syndrome (HUS) attributable to enterohemorrhagic* E. coli *(EHEC) in France (King* et al. *2009). For a color version of this figure, see www.iste.co.uk/membre/microbiological.zip*

An exposure model was built to describe the outbreak of HUS that occurred in France in 2005, covering the food pathway from packaging to consumption (Delignette-Muller *et al.* 2008).The estimated total number of consumed ground beef patties was 2,155. The ingested dose depended on the initial level of contamination (C in CFUs/g), the consumption preference (CP: raw, rare, medium or well-done) and the serving size (S in g). The needed parameters C, CP and S were estimated for each age group. The dose for a specific age group was assumed to follow a Poisson distribution with a parameter specific to each age group equal to C x S_{age} x 10^{-Rage}. R_{age} is the number of the log decimal reduction that depends on the consumption preference C_{Page}. The concentration C was estimated from microbial detection and counts performed on 22 frozen patties sampled from the contaminated batch: C = 5.8 CFUs/g, with 95% credible interval [3.2, 9.4]. The serving size was assumed to follow a log-normal distribution with a log scale mean equal to 4.7 + 1.1 x (1-exp(-0.18 x age) and standard deviation equal to 0.40. The CP proportions were assessed for children over two years old and children two years old and below, and the vectors of proportion for raw, rare, medium and well-done were, respectively (0, 0.02, 0.38, 0.60) and (0.01, 0.19, 0.52, 0.28). For the rare, medium and well-done CPs, the decimal log reductions were assumed to follow gamma distributions with, respectively, the following means and standard deviations: (0.47, 0.25), (0.98, 0.23) and (2.4, 0.46). Finally, the number of consumed ground beef patties for each group was estimated to be between 35 and 39.

The probability of HUS can be derived from the following equation:

$$P(HUS|age) = 1 - \left(1 - p_{age}\right)^{D_{age}}$$

where D_{age} is the ingested dose assessed from the described exposure model and

$$p_{age} = p_0 \times \exp(-k \times age)$$

In order to estimate the two parameters of the dose–response model p_0 and k, a Bayesian framework was used. The model inputs were the distribution of ingested doses per age assessed from the exposure model, the number of exposed children per age, the posterior distribution of the parameter k from the Bayesian analysis of the log-linear Poisson regression model applied to the epidemiological surveillance system and the total number of observed HUS cases during the outbreak (16 confirmed cases of HUS).

The obtained results were $k = -0.38$ with 95% credibility interval [–0.40, –0.36] and $\log10(p_0) = -2.33$ with 95% credibility interval [–2.56, –2.12].

5.3.3. *Combination of data from different outbreaks*

Typically, information from an epidemic survey provides only one data point on a population dose–response curve (the probability of illness for a particular exposure scenario). A collection of outbreak datasets is needed to more accurately and reliably establish the dose–response model. Combining multiple outbreaks provides a broader range of exposure doses and susceptibilities of individuals and contributes to a better estimation of the parameters of the dose–response relationship. Bayesian methods can be used to develop hierarchical models to combine data from different outbreaks, perform meta-analyses and provide estimates by fitting the model simultaneously to data from different outbreaks. The hierarchical framework for estimating dose–response model parameters has been applied to various pathogens by Teunis *et al.* (2008, 2010, 2012) and Thebault *et al.* (2013).

To illustrate this approach, we take the example of the dose–response relationship established by ANSES (Chartier *et al.* 2017) for *Trichinella*, which was built upon a previously published dose–response model (Teunis *et al.* 2012). *Trichinella* is a parasitic roundworm belonging to the class of nematodes. The *Trichinella* genus includes various species and genotypes. From an epidemiological point of view, wildlife is the main reservoir for *Trichinella*. In France, three species have been isolated in animals: *T. spiralis*, *T. britovi* and *T. pseudospiralis*. It should be noted that other species have been isolated in human cases[1].

The specific dose–response model for *Trichinella* is:

$$P(infection|N_F \text{ and } N_M) =$$

$$1 + \frac{\Gamma(\alpha+\beta)\times\Gamma(\beta+N)}{\Gamma(\alpha+\beta+N)} - \frac{\Gamma(\alpha+\beta)\times\Gamma(\beta+N_F)}{\Gamma(\alpha+\beta+N_F)} - \frac{\Gamma(\alpha+\beta)\times\Gamma(\beta+N_M)}{\Gamma(\alpha+\beta+N_M)}$$

1 Data from the Cochin Hospital Center, Laboratoire conventionné avec l'INVS pour la surveillance des cas humains de trichinellose. Available at http://cnrdestrichinella.monsite-orange.fr/ [accessed October 2016].

The data included nine documented outbreaks. The quantity of larvae (dose) ingested by the individuals is unknown but is specified as uncertain in the model with a negative binomial distribution of parameters: $(C \times m\,;\rho)$. Exposure doses are thus characterized by the observed mean concentration of the responsible foods (C), the mean quantity ingested in g (m) and a dispersion parameter (ρ). When parameter ρ increases, the dose distribution approaches a Poisson distribution.

Figure 5.9 graphically illustrates the statistical model for adjusting the α and β parameters of the dose–response relationship.

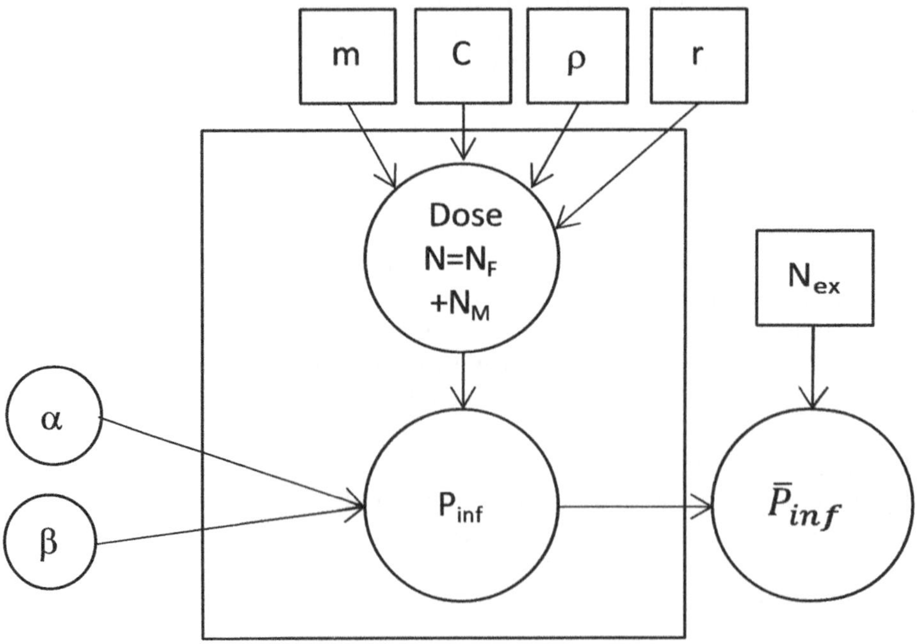

Figure 5.9. *Visualization of the Bayesian statistical model for estimating the* α *and* β *parameters of the dose–response relationship. The squares correspond to fixed components of the model (m: mean ingested quantity of the food, C: mean concentration in the food,* ρ*: dispersion parameter, r: ratio of female to male larvae,* N_{ex}*: number of exposed individuals),* P_{inf} *is the probability of infection given NF and NM (respectively, the numbers of female and male larvae) and* $\bar{P}_{inf}$ *is the number of infected individuals. The circles are random components of the model*

The model can be adjusted thanks to a random sampling set (Markov chain Monte Carlo (MCMC)), with two nested iteration levels.

Level 1 involves parameters α and β. During every iteration, a pair of parameter values is chosen randomly from a non-informative prior distribution. Level 2 involves several doses randomly chosen followed by the calculation of infection probabilities as a function of these doses and the values of the α and β parameters randomly chosen in Level 1. The mean of the infection probabilities is calculated for all of the Level 2 iterations. The likelihood (probability) of the observed number of infected individuals out of all exposed individuals is calculated using the formula for the binomial distribution of parameters: $\bar{P}_{inf}$ and N_{ex}. This likelihood is simply the probability of observing a certain number of infected individuals under the exposure conditions characterized by m, C, ρ and Nex and for specific values of α and β.

At the end of the process, there is a series of calculated likelihoods for a series of possible α and β value pairs randomly selected from their prior distributions. Thanks to Bayes' theorem (calculation of conditional probabilities), it is possible to derive, from this series of likelihood values, the posterior distribution of the α and β parameters: the probabilities of α and β considering the observed number of infected individuals and the exposure conditions. Here, the posterior distribution represents uncertainty about the determination of the dose–response relationship.

The posterior distribution of α and β is determined outbreak by outbreak or for a set of outbreaks, for example, outbreaks involving the same species of *Trichinella*. Given that the estimation procedure is iterative, its convergence needs to be verified. The Gelman and Rubin criterion, which compares within-chain and between-chain variances of posterior distributions, can be used.

The curve in Figure 5.10 shows the relationship between the number of ingested *Trichinella* larvae and the probability of developing trichinellosis. The result is similar to that obtained by Teunis *et al.* (2012). The median probability of occurrence of trichinellosis following exposure to 10 larvae is close to 0.1. Thanks to the outbreak data published since 2012, there is more low-dose information here than in the study by Teunis *et al.* (2012). In terms of goodness of fit, the estimated outbreak attack rates fall within the 95% confidence interval of the prediction. However, for exposure to low doses, the median value tends to overestimate the risk.

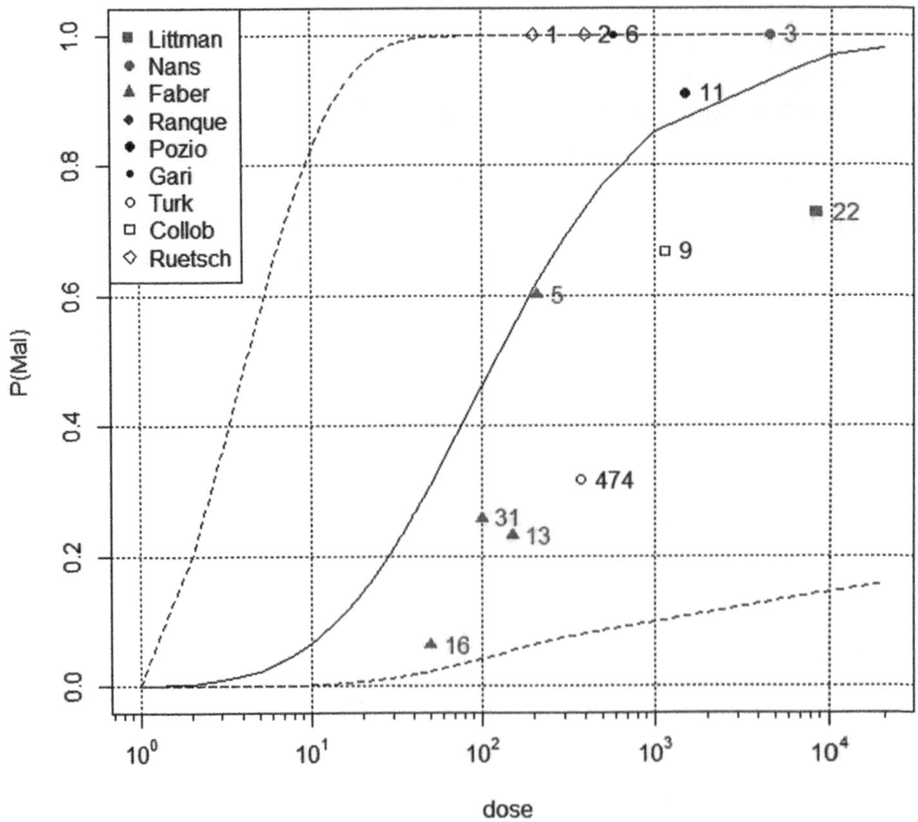

Figure 5.10. *Dose–response relationship estimated based on data collected from nine outbreaks (probability of occurrence of a trichinellosis case as a function of quantities of ingested larvae) (Chartier* et al. *2017). The dotted curves represent the 95% confidence interval for the probability of infection. The dots and numbers represent attack rates as a function of the mean dose and the number of people exposed to the mean dose, respectively. For a color version of this figure, see www.iste.co.uk/membre/microbiological.zip*

5.4. Discussion and conclusion

In microbiology, the models currently used are based on the single-hit model theory and the absence of a threshold. Thus, the notion of a minimal infectious dose is no longer used and has been replaced by infectious doses associated with a level of response, for example, 1% or 10%. It is crucial when establishing dose–response relationships or when using them to characterize risk to distinguish between models estimating a conditional and marginal probabilities of infection/disease. The conditional model applies in principle to an individual: the probability of knowing the quantity of a pathogen ingested (n: discrete number) and the result of the interaction between pathogen and host (p: real number between 0 and 1): P(illness|n,p).

Since it is practically impossible for an individual to determine his or her values of n or p, the conditional model is impossible for these two parameters. The value of n, the number of pathogens actually ingested by an individual, is generally uncertain. However, it is possible to describe the variability of n for a group of individuals through probability distributions. Thus, the model does not give a unique probability of infection for an individual but can give a distribution of it. The mean of the distribution of the possible infection probabilities constitutes what is called the marginal probability, that is, we move from an individual to a group of individuals. In other words, the model expresses the average probability of observing the infection/disease event, which is nothing other than the proportion in the population that will experience the event. The marginal probability also applies to the value of p because of the variability of the performance of the immune systems of individuals in a population and also that of the virulence between strains of the same pathogen. The marginal model thus gives an expression of the probability of infection/disease at the population level. A marginal model for n and p gives the probability of infection/disease for an individual taken at random and for which neither the immune status, the strain to which they could have been exposed nor the quantity n are individually known.

Regardless of the data used, the main objective when fitting dose–response relationships is to estimate the parameter p and, when possible, its distribution. Indeed, at the stage of risk characterization, it is p or its distribution that is used. The question now is whether to estimate p as a single value or to estimate a distribution of p. This depends essentially on the type of data. When we only have experimental data from "volunteers", the group generally consists of homogeneous, healthy people, often males, of very similar ages, who have been exposed to the same strain of the pathogen. It is therefore to be expected that the value of p will not be very variable and, in general, a model is fitted with a constant parameter p. This context makes it difficult to extrapolate the dose–response relationship to other groups of individuals. We face the same situation when we use data from a foodborne outbreak in a nursing home or in a kindergarten. Instead of estimating the distribution of p, we can simply estimate different values of p by stratifying the population by age category, gender or any other factor considered relevant. The estimation of the distribution of p is meaningful when we are able to combine data from several foodborne outbreaks representing different exposure scenarios and involving different strains of the same pathogen and also individuals of different ages, sexes, immune status, etc. Even when it is possible to accumulate data from different outbreaks, however, there remains a challenge in determining the weights to be given to each of the outbreaks for the estimation of a representative p distribution in the target populations. It is widely accepted that the majority of foodborne cases remain sporadic and are not covered by outbreaks. We are not sure that the strains or individuals involved in outbreaks are comparable to those implicated in the sporadic cases.

Although the hierarchical framework for estimating dose–response relationships from combined outbreak datasets is very attractive, the results obtained are still difficult to use. Indeed, the large uncertainties about the parameters of the dose–response model illustrated in the Bayesian approach by the posterior distributions do not allow precise prediction of the probability of infection or illness. The high uncertainty may be primarily attributed to the difficulty in accurately estimating the actual doses ingested by the exposed individuals and the fact that the well-documented outbreaks concern specific groups of host individuals that are not necessarily representative of the target population. In practice, the available points of observation (attack rate, dose) may correspond to groups of host individuals with very different susceptibilities, which means that the exposure doses can no longer be considered independent of host susceptibility, as is the case in experimental challenge trials. In other words, the combination of data from outbreaks that occurred in different settings, such as nursing homes, primary schools and hospitals, will without doubt bias the dose–response assessment. A high attack rate may be observed because of exposure to high doses or because the exposed people belong to a highly susceptible group with severe pre-conditions.

For all these reasons, dose–response models that include a p distribution should be used with great caution. We should make sure that the distributions are not biased and that they correspond to the variability that we seek to consider in the risk assessment.

Taking the example of the beta distribution when the shape of the distribution is symmetric or when its variance is moderate, the exponential model is a good approximation of the beta-Poisson model. This observation can be used to adopt a stratification strategy for the population and fit exponential models by strata.

5.5. References

Black, R.E., Levine, M.M., Clements, M.L., Hughes, T.P., Blaser, M.J. (1988). Experimental *Campylobacter jejuni* infection in humans. *Journal of Infectious Diseases*, 157(3), 472–479.

Chartier, C., Casabianca, F., Vallée, I., Villena, I. (2017). Avis du 16 décembre 2016 révisé le 14 mars 2017 de l'Agence nationale de sécurité sanitaire de l'alimentation, de l'environnement et du travail relatif à la contamination de produits de charcuterie crue par *Trichinella* spp.

Clyde, M. (2018). BAS: Bayesian variable selection and model averaging using Bayesian adaptive sampling. R documentation.

Delignette-Muller, M.L., Cornu, M., AFSSA STEC Study Group (2008). Quantitative risk assessment for *Escherichia coli* O157:H7 in frozen ground beef patties consumed by young children in French households. *International Journal of Food Microbiology*, 128(1), 158–164.

Goulet, V., Jacquet, C., Martin, P., Vaillant, V., Laurent, E., de Valk, H. (2012). Surveillance de la listériose humaine en France, 1999–2011. *Bulletin épidémiologique hebdomadaire*, 50, 47–50.

Haas, C.N. (2002). Conditional dose-response relationships for microorganisms: Development and application. *Risk Analysis*, 22(3), 455–463.

King, LA., Mailles, A., Mariani-Kurkdjian, P., Vernozy-Rozand, C., Montet, M.P., Grimont, F., Pihier, N., Devalk, H., Perret, F., Bingen, E., Espié, E., Vaillant, V. (2009). Community-wide outbreak of *Escherichia coli* O157:H7 associated with consumption of frozen beef burgers. *Epidemiology and Infection*, 137(6), 889–896.

Leclercq, A., Kooh, P., Augustin, J.-C., Guillier, L., Thébault, A., Cadavez, V., Gonzales-Barron, U., Sanaa, M. (2020). Risk factors for sporadic listeriosis: A systematic review and meta-analysis. *Microbial Risk Analysis*, 100128.

Mazucheli, J., Menezes, A.F.B., Nadarajah, S. (2017). Mle.tools: An R package for maximum likelihood bias correction. *The R Journal*, 9(2), 268.

Medema, G.J., Teunis, P.F.M., Havelaar, A.H., Haas, C.N. (1996). Assessment of the dose–response relationship of *Campylobacter jejuni*. *International Journal of Food Microbiology*, 30(1–2), 101–111.

Pouillot, R., Hoelzer, K., Chen, Y., Dennis, S.B. (2015). *Listeria monocytogenes* dose response revisited – Incorporating adjustments for variability in strain virulence and host susceptibility. *Risk Analysis*, 35(1), 90–108.

Teunis, P.F.M., Ogden, I.D., Strachan, N.J.C. (2008). Hierarchical dose response of *E. coli* O157:H7 from human outbreaks incorporating heterogeneity in exposure. *Epidemiology and Infection*, 136(6), 761–770.

Teunis, P.F.M., Kasuga, F., Fazil, A., Ogden, I.D., Rotariu, O., Strachan, N.J.C. (2010). Dose–response modeling of *Salmonella* using outbreak data. *International Journal of Food Microbiology*, 144(2), 243–249.

Teunis, P.F.M., Koningstein, M., Takumi, K., Van Der Giessen, J.W.B (2012). Human beings are highly susceptible to low doses of *Trichinella* spp. *Epidemiology and Infection*, 140(2), 210–218.

Thebault, A., Teunis, P.F.M., Le Pendu, J., Le Guyader, F.S., Denis, J.-B. (2013). Infectivity of GI and GII noroviruses established from oyster related outbreaks. *Epidemics*, 5(2), 98–110.

Xie, G., Roiko, A., Stratton, H., Lemckert, C., Dunn, P.K., Mengersen, K. (2017). Guidelines for use of the approximate beta-Poisson dose–response model. *Risk Analysis*, 37(7), 1388–1402.

Zelterman, D. (2005). *Discrete Distributions: Applications in the Health Sciences*. John Wiley & Sons, Hoboken.

6

A Guide for Microbiological Risk Characterization

Laurent GUILLIER and Moez SANAA
ANSES, Department of Risk Assessment, Maison-Alfort, France

6.1. Introduction

Risk characterization is defined as "The qualitative and/or quantitative estimation, including attendant uncertainties, of the probability of occurrence and severity of known or potential adverse health effects in a given population based on hazard identification, hazard characterization and exposure assessment" (Haberbeck *et al.* 2018; Codex Alimentarius Commission 1999).

Risk characterization is the final step in the quantitative risk assessment process. It thus integrates the findings from the three previous components presented in Chapters 1, 2, 3, 4 and 5.

The risk assessment process is initiated by a question from a risk manager. The output of the risk characterization completes the quantitative risk assessment process that has been proposed to bring responses to the risk manager. The output of the risk characterization can subsequently be used to make appropriate risk management decisions.

Concretely, the result of the process is the "risk estimate". The risk characterization can often include one or more estimates of risk, and usually proposes evaluations of risk management options. Those estimates can take several forms (risk per portion, risk for a population, etc.) and may include economic and other evaluations in addition to estimates of risk attributable to the management options.

The risk characterization presented in this chapter corresponds to the classical "bottom-up" approach. Risk estimates can also be assessed through top-down approaches using observational epidemiological information (FAO and WHO 2020). Source attribution models can also be used to assess risk based on methods combining both "top-down" and "bottom-up" approaches (Mughini-Gras *et al.* 2019).

Risk characterization differs according to the methods of risk assessment used: quantitative, semi-quantitative or qualitative. The choice of the most appropriate method depends on the availability and quality of data, the expertise and the time available to produce risk estimates. After briefly describing qualitative and semi-quantitative methods, this chapter focuses on the quantitative risk assessment method, which is more widely used.

6.2. Qualitative and semi-quantitative risk characterization

6.2.1. *Qualitative risk characterization*

In qualitative risk characterization, the output is generally descriptive and is not directly tied to a quantified measure of risk. This requires an overall textual estimate of the risk. A qualitative risk characterization should clearly show how the qualitative conclusion on the risk estimate has been reached, with narrative text that provides the available knowledge for hazard identification and characterization and for exposure assessment. The qualitative risk characterization qualification can also be obtained by combining qualitative assessments of exposure ("low", "very", etc.) and hazard characterization ("minor", "moderate", "severe", etc.). A risk matrix should then be used to combine categorical labels. This risk matrix provides qualification of the risk at the intersection of likelihood and severity. Table 6.1 illustrates a possible risk matrix.

6.2.2. *Semi-quantitative risk characterization*

The semi-quantitative risk assessment method is somewhere between the textual evaluation of risk in qualitative risk assessment and the evaluation in quantitative risk assessment with a score. It offers a more consistent and rigorous approach for assessing and comparing risks and risk management strategies than qualitative risk assessment. This method avoids some of the ambiguities that a qualitative risk assessment may produce. It requires far less data and far fewer models than quantitative risk assessment. Semi-quantitative risk assessment involves assigning numbers to qualitative estimates of exposure and the dose–response relationship in the form of probability ranges, weights or scores. These are then combined through a mathematical operation to get a risk estimate, which is usually a scale.

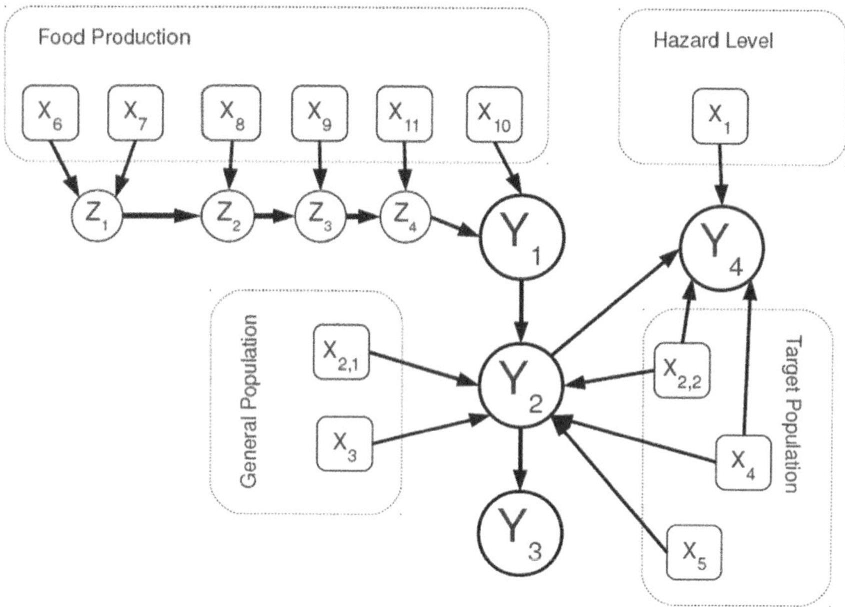

$$Y_1 = min(1, \max(X_6 X_7, X_8) \times X_9 X_{10} X_{11})$$
$$Y_2 = min(1, Y_1 X_{2,1} X_3)$$
$$Y_3 = 365 \times Y_2 X_{2,2} X_4 X_5$$
$$Y_4 = 100 + \log_{10}(Y_2 X_{2,2} X_4 X_1)/17.56$$

Figure 6.1. *Relation between the 11 inputs (X) and the outputs (Y) of the Risk Ranger tool. From Guillier* et al. *(2013)*

Figure 6.1 illustrates one of the tools used for semi-quantitative risk assessment, the Risk Ranger tool. The main output is Y4. It is scaled logarithmically between 0 and 100 (where 0 represents the risk probability of illness for a mild hazard of less than one case per 10 billion people per 100 years).

6.3. Quantitative measures of risk

6.3.1. *Measures of risk*

The public health risk upon exposure to a contaminated food can be expressed in several ways. Figure 6.2 illustrates the different ways of expressing risk.

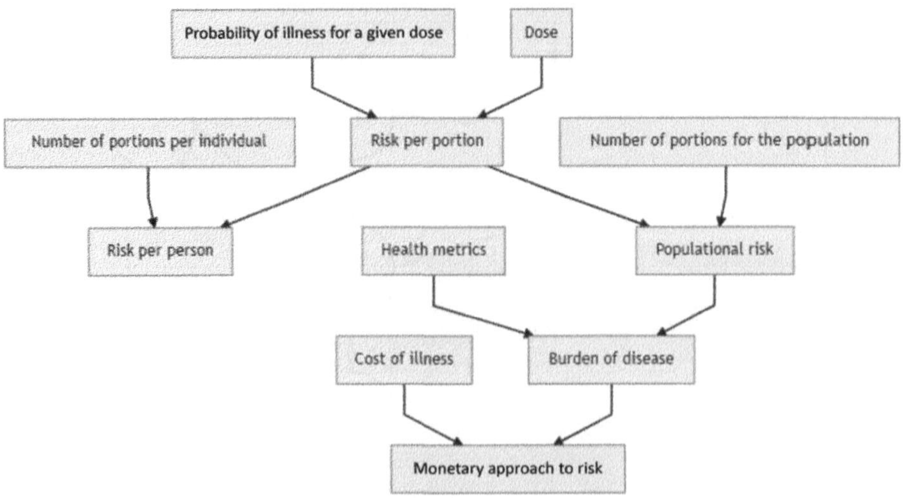

Figure 6.2. *Different measures of risk*

The simplest way to characterize risk is to calculate the risk per portion. In this situation, risk is simply obtained by combining the exposure of the consumer expressed in cells per portion with the dose response that provides the probability of an adverse effect for a given dose. The resulting risk is expressed as a probability of illness per portion. The second type of output is the characterization of the risk for each individual in a given period of time (usually a year). The risk is thus obtained by combining the probability of illness per portion with the number of portions consumed by individuals. It is expressed as a probability of illness per year. A third type, and the far most common, is the risk characterization at the population level, that is, the number of illnesses to be expected from the consumption of a particular food item among a population in a specific area and time period. Over the last several years, other risk outputs have been proposed (see sections 6.2.2 and 6.2.3).

Acknowledging that risk estimates are prone to uncertainty and bias, risk assessments often do not aim to estimate the absolute risk and public health impact, but are rather used as a tool to assess the relative impact of different risk mitigation strategies. Once the "baseline" risk assessment model is constructed, different scenarios (new interventions on the food chain, introduction of microbiological criteria, change in consumer practice, etc.) can be evaluated and their relative impacts on illness can be calculated (Codex Alimentarius Commission 1999).

The output of the risk characterization can be expressed as the relative risk:

$$relative\ risk = \frac{number\ of\ cases\ after\ intervention}{number\ of\ cases\ for\ baseline\ scenario}$$

or as a risk reduction

$$risk\ reduction\ (\%) = 100 \cdot \left(1 - \frac{number\ of\ cases\ after\ intervention}{number\ of\ cases\ for\ baseline\ scenario}\right).$$

6.3.2. *From risk to burden of disease*

In order to go beyond the number of disease-related cases saved with interventions, two concepts are generally used: quality-adjusted life years (QALYs) and disability-adjusted life years (DALYs). QALYs correspond to the life years gained with an intervention, taking into account the quality of life during these years. A QALY is assigned a coefficient of 1 if it is a healthy year and less than 1 if the quality of life is reduced. These coefficients are based on population surveys.

The DALY metric quantifies the gap between a life lived in perfect health and the current health status as the number of healthy life years lost due to illness (as years lived with disability (YLDs)) and to premature death (years of life lost (YLLs)). The calculation of the YLLs takes into account the numbers of deaths in a reference year for a specific cause of disease (d) and the remaining life expectancy at age of death (le) within each group of the population (i from 1 to n):

$$YLL = \sum_{i}^{n} d_i \cdot le_i$$

To estimate YLDs for a population, the number of human cases with disabilities (I) is multiplied by the average duration of the disease (L) and a weight factor (DW) that reflects the severity of the disease:

$$YLD = I \cdot DW \cdot L$$

Figure 6.3 illustrates these two concepts.

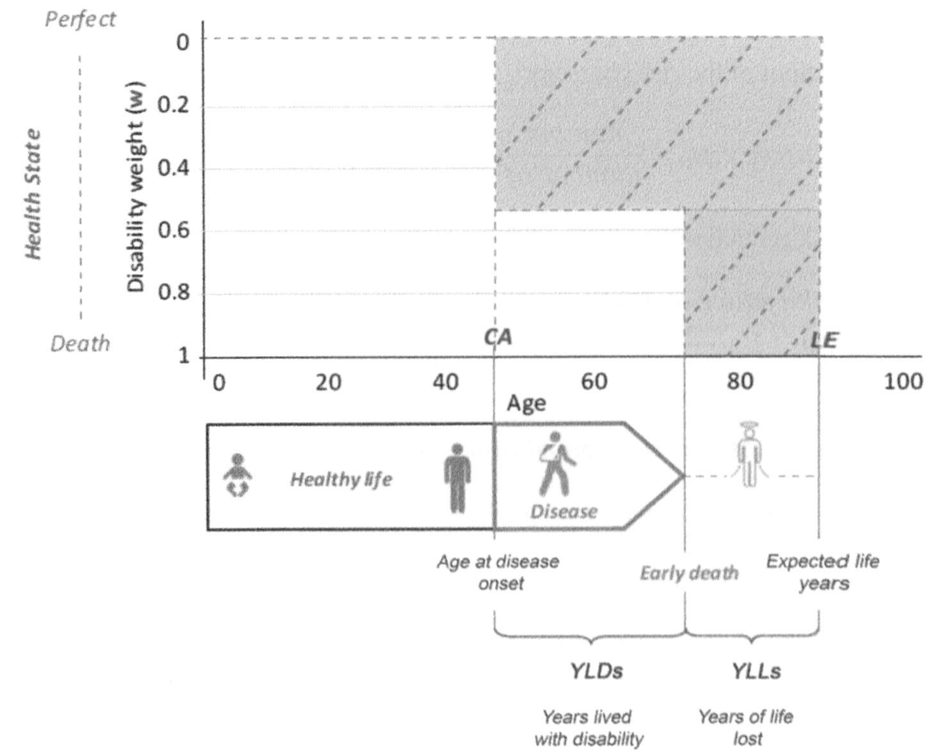

Figure 6.3. *Illustration of DALYs with the case of a person who falls sick and dies after a period of disability (Boué 2017). For a color version of this figure, see www.iste.co.uk/membre/microbiological.zip*

Although QALYs and DALYs rely on the same conceptual framework, they are not totally equivalent, as they are partly based on different methods for eliciting quality of life and disability scores (Sassi 2006). Both indicators have been used for assessing the burden of disease for microbiological hazards (Mangen *et al.* 2013; Daniel *et al.* 2020). In 2018, The Burden of Communicable Diseases in Europe (BCoDE) study estimated the DALYs for 31 selected diseases in the European Union (EU) (Cassini *et al.* 2018). In this context, DALYs seemed to be preferred to QALYs in the context of microbial risk assessment.

The use of such metrics allows us to rank the burden associated with different hazards (Augustin *et al.* 2020) or measure the burden of a food category caused by several microbiological hazards (De Oliveira Mota *et al.* 2020). These metrics are also essential in risk benefit assessments (Membré *et al.* 2020). An example of estimating risks and quantifying benefits using the DALY metric has been recently proposed (De Oliveira Mota *et al.* 2019, 2020). This study estimated that red meat consumption in France contributes to a mean of 19 DALYs per 100,000 people per year for cases of colorectal cancer, 21 DALYs for cardiovascular disease and 6.6 DALYs for foodborne pathogens. These DALYs associated with risk were further compared to gain the DALYs associated with meat iron intake.

DALYs also provide the basis for cost–benefit analysis (Duret *et al.* 2019; Suijkerbuijk *et al.* 2019), as explained below.

6.3.3. *Monetary approaches to risk*

6.3.3.1. *Why integrate cost in risk characterization?*

The characterization of the health impacts of different microbiological hazards (e.g. incidence, mortality rates) is not the only criteria to be taken into account to help the risk manager prioritize interventions. Other dimensions must also be taken into account to optimize the allocation of the means of control and the commitment to intervene in risk reduction strategies.

Two purposes are generally identified for costs. The first is the prioritization of hazards based on the costs associated with different diseases. Decision-makers can indeed consider prioritizing action for those situations that are most costly to civil society. From this perspective, cost evaluation makes it possible to answer the question: which microbiological hazards or which food/hazard pairs cost society the most?

The second use of costs is related to measuring the effectiveness of the means allocated by risk managers to reduce the health burden. It is possible that a disease X causes a higher health burden than a disease Y, but that for the same public expenditure (a consumer communication campaign, a control plan for a sector, the introduction of a regulation, etc.), the health gains for Y are higher than for X. To take account of this differentiated impact of public investment, we can try to compare the cost–effectiveness ratio of the various possible interventions and determine which of a range of possible actions should be given priority. The results of the analysis can be expressed, for example, in the number of life years gained per euro invested. These "cost–effectiveness" approaches are also used to identify among all levels of a food production chain (primary producers, processors, distributors, etc.) which actions should be prioritized (see, for example, Havelaar *et al.* (2007) and Van Wagenberg *et al.* (2016)). Cost–benefit analysis (Bergion *et al.* 2018) comes close to these cost–benefit approaches. It can integrate the economic value of health benefits in order to answer the following question: for a given disease, to what extent is the economic value of the health gains induced by an intervention greater or less than the associated costs?

6.3.3.2. *How can the cost of illness be calculated?*

One means for the assessment of the economic burden of illness is to measure the loss of "well-being" that it causes in society as a whole. The first component concerns the economic evaluation of impacts in terms of mortality. Mortality is understood through the notion of the "value of statistical life" (VSL), which expresses the amount that society is willing to pay for an action that reduces the risk of death by 1/1,000 in a population of 1,000 individuals and thus results in avoiding premature death (= one statistical life) (Charpentier and Cherrier 2019). This notion should not be interpreted as an economic "value" given to human life. It rather reflects a willingness to pay more for a situation that provides safer and better health. The values obtained vary from country to country and recommendations have been made by international organizations (Lindhjem *et al.* 2011). As an example, in France, a value of €3 million has been proposed (Téhard *et al.* 2020).

The second component concerns the variation in well-being associated with a non-fatal disease, and therefore the value of society's willingness to pay for a reduction in the non-lethal damage induced by this disease. It is generally broken down into three types of costs. The direct costs correspond to the expenses induced by the occurrence of the disease. They are directly quantifiable in monetary units and may correspond to expenditure (hospitalization costs, medicines, patient transport, etc.). The indirect costs do not give rise to monetary expenditure. They correspond to the economic value of the losses due to the time that could not be spent on other activities by patients or their relatives (e.g. parents of sick children). These costs usually correspond to production loss due to absenteeism. The intangible costs correspond to the loss of quality of life (e.g. pain, disability, after-effects) resulting from the illness, either temporarily or permanently.

6.3.4. *Toward other approaches*

Beyond the use of health metrics and costs, other metrics should be taken into account by risk managers. Taking decisions is complicated as, for each intervention scenario, along with risk reduction, other outcomes can be estimated (Guillier *et al.* 2016). Along with risk, the nutritional, organoleptic and environmental qualities may be in tension (Prache *et al.* 2020). In the same way, the interests of the different stakeholders (primary production, retailers, consumers, etc.) of the food chain might strongly differ. This underlines the need to seek ways of compromising between different properties and between different actors and to have the tools to do so. In this context, there has been renewed interest in the multi-criteria decision analysis (MCDA) methodology, which can help a decision-maker to make a choice in a multidimensional environment (Duret *et al.* 2019; Garre *et al.* 2020). MCDA can help to overcome issues involving conflicting and non-comparable outcomes based on a formalized decision-making process that determines the best compromise, according to preferences established by the decision-maker. Their role is essential because they formalize the weightings and compensations granted to the different criteria.

6.4. From results of risk assessment to the application of measures in the food chain

The objective of the risk manager is often to use the results of the quantitative risk assessment to improve the safety of a food product. In order to help implement the control measure in the real context of the food production chain, risk-based metrics have been proposed. These metrics provide a convenient way of relating food safety management and food processing operations (good manufacturing practices (GMPs) and good hygiene practices (GHPs), an HACCP program, etc.) to risk.

6.4.1. *Appropriate level of protection*

The EU agreement on sanitary and phytosanitary (SPS) measures (WTO 1995) points out that Member States have the right to adopt SPS measures to achieve their (self-determined) health protection level. This level is called appropriate level of protection (ALOP). The ALOP is estimated by the country establishing a sanitary measure to protect the lives or the health of humans within its territory. In the context of food safety, an ALOP is a statement of the degree of public health protection that is to be achieved by the food safety systems implemented in a country. The ALOP can be stated in a range of terms (implicit ALOP), such as "maintain the actual level of protection", or as a quantitative expression of the probability or incidence of disease (explicit ALOP) (Tuominen *et al.* 2007). It is typically expressed in terms relevant to public health, such as the number of cases per 100,000 population per year. Two different situations can be distinguished in setting a quantitative ALOP.

Typically, an ALOP would be articulated as a statement related to the disease burden associated with a particular hazard–food combination and its consumption in a country (FAO and WHO 2020). In these cases, the *status quo* of public health protection in relation to a disease can be the starting point for defining an ALOP (Gkogka *et al.* 2013a), as the ALOP can represent "*the current public health status in relation to food safety*" (EFSA 2007). Defining the ALOP in this context is still a challenge as the real incidence of a disease and the actual fraction of cases attributable to a food are difficult to assess (Gkogka *et al.* 2013b). The ALOP can then be derived from source attribution methods or through the output of the risk assessment baseline scenario (Gkogka *et al.* 2020).

The ALOP can also be used in the context of continual improvement in relationship to disease reduction. An example of data-driven national objectives aiming to improve health can be found in the Healthy People 2030 program in the United States of America. For each foodborne pathogen, the current or past cases of laboratory-confirmed infections per population of 100,000 per year are reported and the target objective for 2030 is proposed (e.g. for campylobacteriosis, the objective is to achieve a reduction from 12.7 to 8.5 cases per 100,000 by the 2030 horizon).

6.4.2. *Food safety objective*

The ALOP is mainly used by risk managers with the aim of informing the public. However, the ALOP may not be the most adequate concept to develop and implant the necessary control measures throughout the farm-to-fork pathway (Havelaar *et al.* 2004). The terms in which the ALOP is expressed (e.g. incidence) do not afford a tangible criterion that the industry or the other operators of the food chain can use for food safety management (Gorris 2005). In this context, the ICMSF (2002) proposed the creation of a new concept, the food safety objective (FSO). They defined an FSO as "the maximum frequency and/or concentration of a hazard in a food at the time of consumption that provides or contributes to the ALOP". The FSO is more straightforward and easier to understand for industrial actors and other operators of the food chain. The FSO can thus be used as a starting criterion for selecting and developing their control measures.

According to the ICMSF (2002), FSOs can be derived from ALOPs by using the dose–response curve in which the probability of the adverse effect is represented as a function of the ingested dose. The FSO according to this method is a simple inverse reading of the dose–response curve. However, using only a dose–response curve, we cannot assess the probability of the adverse health effect for several exposures nor the risk for the overall population but only the risk for one single exposure (which is not the usual way to define an ALOP). Havelaar *et al.* (2004) and Rieu *et al.* (2007) have thus recommended taking into account the frequency and amount of food consumption, the prevalence of contaminated food and the dose–response model to derive the FSO.

6.4.3. *Performance objective/performance criterion*

Two other risk-based metrics proposed by the FAO and WHO have been defined (2004): the "performance objective" (PO), defined as "the maximum frequency and/or concentration of a hazard in a food at a specified step in the food chain before the time of consumption, that provides or contributes to an FSO or ALOP, as applicable", and the "performance criterion" (PC), defined as "the effect in frequency and/or concentration of a hazard in a food that must be achieved by the application of one or more control measures to provide or contribute to a PO or an FSO".

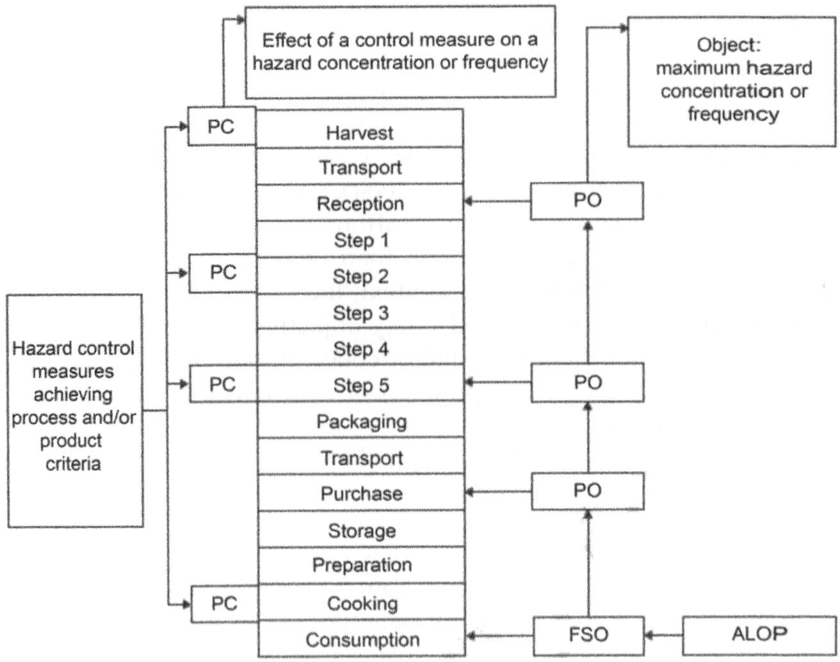

Figure 6.4. *Risk management metrics in the food production chain. ALOP: appropriate level of protection, FSO: food safety objective, PO: performance objective, PC: performance criterion*

6.4.4. *Process and product criteria*

The performance objective and performance criterion must also be translated into terms and concepts that food operators can introduce in their food management system, the so-called "process criteria" and "product criteria" (Membré and Boué 2018). Van Schothorst (2002) defined these criteria as the control parameters (such as time, temperature, water activity, pH, etc.) that can be applied at a specific step to reach an efficiency criteria. In an HACCP context, these would correspond to the critical control limits of a process and would be defined as the parameters of a food product that are essential to ensure that an FSO is reached (Van Schothorst 2002, 2005). This set of objectives, criteria and limits can be considered as part of HACCP programs and GMP/GHP guides in order to finally achieve an FSO (Van Schothorst 2005). The sequential nature of the food process steps and the links with the above-mentioned food safety metrics are shown in Figure 6.4.

6.5. Concepts of variability and uncertainty in the context of risk characterization

Uncertainty and variability are important concepts in risk characterization. While it is recommended to consider variability and uncertainty separately (Codex Alimentarius Commission 1999), these two concepts are nevertheless frequently misunderstood.

6.5.1. *Variability*

Variability refers to the inherent heterogeneity of data used in a quantitative risk assessment. Variability is a quantitative description of the range or spread of a dataset and is either expressed through statistical metrics (e.g. standard deviation, variance, quantiles or interquartile ranges) or modeled with statistical distributions (normal, exponential, Gamma, triangular, etc.). The components of a risk assessment (i.e. of the hazard characterization and the exposure assessment steps) are usually variable. Consequently, it leads to a highly variable output. Several examples of variability have been described in other chapters. In the context of hazard characterization, variability may involve individual susceptibility or strain virulence. For the exposure assessment step, numerous inputs are variable, such as the initial contamination, the ability of strains to grow in a food, the conditions used for food preparation and storage, and the serving size.

The probability of observing an adverse effect following the consumption of a product thus varies extensively from serving to serving, from individual to individual, from day to day and from region to region, etc.

6.5.2. *Uncertainty*

Just as variability is everywhere in a risk assessment model, so too is uncertainty. Uncertainty refers to a lack of data or an incomplete understanding of the phenomenon during modeling. Three types of uncertainty are usually defined (i) scenario uncertainty, (ii) model uncertainty and (iii) parameter uncertainty.

Scenario uncertainty is linked to our representation of the situation that conducts risk. Errors in the selection of exposure routes, populations of concern or models can be made. By definition, as these uncertainties are linked to an incomplete knowledge of the real situation, they are usually not identified and are therefore difficult to quantify. In the domain of quantitative microbial risk assessment, a scenario uncertainty frequently arises when considering only the "baseline" situation, while the illnesses are not linked to the extreme percentiles of this baseline situation but to a completely different situation that is not, and cannot be, predicted by the model. An example could be the O104:H4 outbreak linked to the consumption of sprouts (Karch *et al.* 2012). Before this outbreak, whatever the quality of the exposure model for sprouts or the classical dose response used for stx pathogenic *E. coli*, no one could have foreseen this scenario, with a particular combination of virulence factors and such a high number of hemolytic and uremic syndromes and deaths in the adult population. Such unpredictable events are referred to as black swan events and, despite the impossibility of foreseeing their occurrence, knowledge remains the key to anticipating them (Flage and Aven 2015; Koch *et al.* 2020).

Model uncertainties arise because of gaps in the background scientific knowledge that is required to make predictions. At least three types of model uncertainty can be identified in quantitative risk assessment (Pouillot and Guillier 2020). The first type includes situations where there are oversimplified representations of reality. It is quite common in exposure assessment to simplify the models used to predict the exposure. For example, the lag time before the growth of pathogens and the microbiological competition with the background microflora are very often omitted (EFSA Panel on Biological Hazards *et al.* 2018). The second type involves the failure to account for correlations between input variables of a risk model. For example, consumer practices are a key element when characterizing exposure. For pathogens to be able to grow during the storage of a fresh food, exposure depends on the temperature of the refrigerators and shelf-life compliance. As the data related to the temperature of the fridge and the shelf-life compliance are usually derived from different studies, their correlation cannot be established. This is a potential source of bias as risk appears to be very sensitive to the proportion of foods stored at high temperatures and after the use-by date (Duret *et al.* 2014). The third type of model uncertainty is related to the use of surrogate data. The presence of knowledge gaps in the exposure assessment or hazard characterization steps is the norm. In the face of such gaps, it is common practice to replace the missing information with existing data. In the case of hazard characterization, because of the impossibility or difficulty of obtaining dose reponse in humans, surrogate data obtained from animals are used (Sanaa 2020). For the establishment of dose response when no data are available for a pathogen, the data obtained with a close but distinct hazard are used (Strachan *et al.* 2005). In exposure assessment, the use of predictive models established for surrogate microorganisms is a common practice (Bartsch *et al.* 2019). These three types of model uncertainties are difficult to quantify and sometimes difficult to identify. These uncertainties should be listed in the document that describes the risk assessment in a specific part (e.g. "review of modeling choices, assumptions and data gaps").

The last type of uncertainty is the focus of risk assessors. It relies on parameter uncertainty. Risk assessments represent reality through mathematical models. Uncertainty in parameter estimates originates from various sources. Sources of uncertainty arise from measurement errors (e.g. for bacterial enumeration), sample representativeness, random errors, etc. Some of these uncertainties may be quantified and are thus integrated in the expression of risk characterization. This type of uncertainty can be reduced with more knowledge. A simple example, for uncertainty in parameters, is the confidence in the mean microbial contamination of a given product. Sampling more samples will not reduce the variability of the contamination from sample to sample, but it will help to obtain a better estimate of the variability distribution and its associated statistics (mean, median, variance, etc.).

6.6. Risk characterization by taking into account variability and uncertainty

Considering variability and uncertainty separately is an international recommendation. It is important to evaluate the impact of these two elements. The importance of separating these two concepts has been demonstrated, as mixing them can lead to biased estimates (Nauta 2000).

6.6.1. *Integration of variability*

6.6.1.1. *Characterization of variability of inputs of the risk assessment model*

Characterizing the distribution of a given input of a quantitative microbiological risk assessment (QMRA) model requires theoretical knowledge or observed data collected from this distribution. All the possible values of an input are of importance in the context of risk assessment (and notably the extreme observations).

Three situations can be distinguished. First, the parametric distribution to use for the input may be known (normal, lognormal, gamma, etc.) because of background knowledge on the distribution of the variable of interest. In that case, the parameters of this distribution have to be inferred from the observed data. Second, if there is no specific knowledge of the underlying distribution, various parametric distributions can be fitted to the observed data, and the one that best fits the data and that is reasonable with regard to the data can be chosen. Finally, it is possible to sample directly from observed data (empirical distribution).

6.6.1.2. *Monte-Carlo simulation to integrate variability*

The mean risk per serving is the essential estimate of the risk characterization step. It can be obtained by calculations, but this relies on a complex integration of distributions that makes the calculation almost impossible. It is worth noting that the mean risk per serving cannot be obtained either by using a deterministic approach or with single values of input variables (mean, median, 75th or 90th percentiles) (EFSA Panel on Biological Hazards 2015). The solution to cope with this problem is to use a Monte-Carlo simulation to evaluate this mean risk. The basic principle is to simulate the distribution (from serving to serving) of the risk, and when the number of simulated values tends to infinity, the mean of the resulting values tends to the actual mean that would be obtained mathematically.

The Monte-Carlo method is straightforward to understand in terms of each iteration of the simulation being "one possible serving." At each simulation, whether the consumer is sick or not (binary 0/1 output) following the ingestion of this dose is not determined. Actually, risk assessors do not simulate illnesses but just the probability of illness in order to estimate the arithmetic mean of the risk. The mean risk can be evaluated from the proportion of contaminated servings (proportion of dose > 0, let us call it p) and the mean risk resulting from contaminated servings;

$$mean\,risk = p \cdot \text{mean risk}|\text{dose} > 0$$

It is important to note that if the output obtained by the Monte-Carlo simulation is a distribution of the risks per portion, the usual result is an integration of the variability. The output of this integration results in a single value (the expected number of cases of illness in the considered population). The result of a risk assessment that does not consider uncertainty is thus a single value.

6.6.2. *Integration of uncertainty*

6.6.2.1. *Characterization of the uncertainty of the inputs of the risk assessment model*

Characterization of the uncertainty of each input of a risk model and assessment of their impacts on the output is part of the formal process of uncertainty analysis. Uncertainty characterization of inputs is usually done together with variability characterization. As not all "individuals" for an input have been sampled, there will be some uncertainty around the variability distribution specified. A part of this parameter uncertainty can be estimated using statistical procedures, such as Bayesian inferences and (parametric and non-parametric) bootstrapping (Pouillot and Delignette-Muller 2010). At the end of the process, a distribution of hyper-parameters reflecting their uncertainty is obtained from raw data. The variability of the parameter of interest is then derived conditionally from a given set of uncertain parameters.

6.6.2.2. *Uncertainty of risk characterization*

The integration of uncertainty for the model output depends on the number of uncertain inputs. If there is a limited number of uncertain parameters (with a limited number of possible values), or if there is a limited number of alternative scenarios to be tested, a simple way to consider uncertainty is to run the Monte-Carlo integration of the variability dimension once for each uncertain value/scenario.

When there is a large number of uncertain values/scenarios, the second-order Monte-Carlo simulation method is used. The Monte-Carlo simulation in the variability dimension provides, conditionally from a set of variable input parameters to uncertain parameters, one value for the estimate. It can be written as $Y = f(X)$, where Y is the estimate and X is a set of parameters considered in this simulation as being known exactly (there is no uncertainty within a variability Monte-Carlo simulation). In order to estimate how the estimation of Y changes according to other sets of X, a Monte-Carlo simulation can be derived. The "variability" Monte-Carlo simulation is then embedded within an "uncertainty" Monte-Carlo simulation (Figure 6.5) (Pouillot and Delignette-Muller 2010).

The final result is then an uncertainty distribution of the expected number of cases. Using the median (or the mean) of this distribution provides a central estimate of the expected number of cases; the 2.5th and the 97.5th percentiles (or other levels) of this distribution provide a "credible interval" around this central estimate.

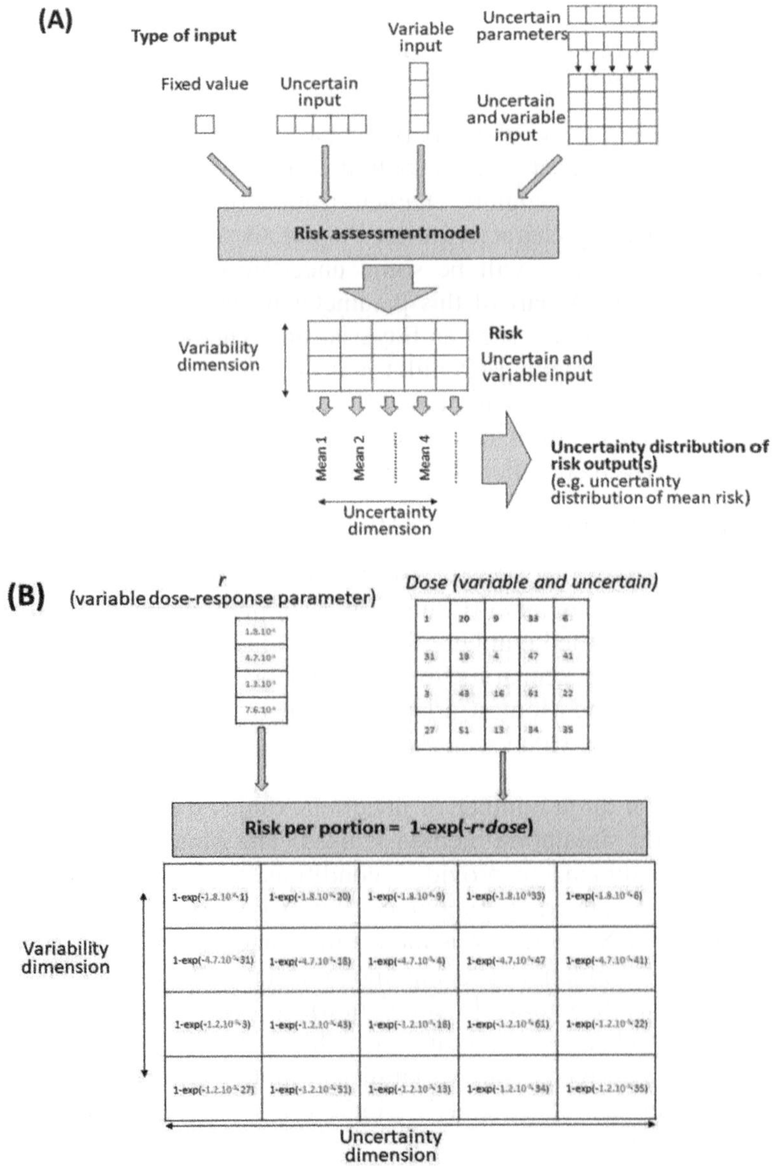

Figure 6.5. *(A) Illustration of the two-dimensional Monte-Carlo simulation with various input types. (B) Practical example of a combination of variability and uncertainty. For a color version of this figure, see www.iste.co.uk/membre/microbiological.zip*

6.7. Software tools and good practices for risk characterization

6.7.1. *Software tools for risk characterization*

Numerous modeling tools have been developed for modeling the growth and inactivation of microorganisms (Tenenhaus-Aziza and Ellouze 2015). They can be used in the context of a risk assessment, but they are not designed for it.

A QMRA usually requires the development of complex mathematical models taking into account the variability and uncertainty of inputs. Risk assessors generally develop their own models implemented in dedicated software, such as @Risk® (Palisade Corporation, NewYork, USA) or the mc2d package (Pouillot and Delignette-Muller 2010) in R® (R CoreTeam), to run simulations and perform Monte-Carlo integrations.

Besides these software primarily fitted for professional risk assessors, software designed for non-specialist users, such as food business operators or food safety authorities, have also been developed. They describe specific or generic processes from the raw material to the end product (exposure assessment modules), input sometimes implemented in a dose–response model assessing the risk of illness (risk assessment modules).

Two open-access tools are worth mentioning. The FDA iRisk tool (Chen *et al.* 2013) is a Web-based system (https://irisk.foodrisk.org/) designed to analyze data concerning microbial and chemical hazards in food and return an estimate of the resulting health burden at the population level. Version 4.0 and above allows the user to perform second-order Monte-Carlo simulations to consider variability and uncertainty separately. The other tool is MicroHibro (González *et al.* 2019; http://www.microhibro.com). It can be used as an online predictive microbiology software. MicroHibro is built on a comprehensive predictive microbiology model database. Probabilistic and point-estimate microbial risk assessment models can be performed.

6.7.2. *Validation and quality control of a risk assessment*

The validity of a risk assessment is based on the relevance of the model structure, its input data, the different assumptions and the interpretation of results. Therefore, quality assurance is a crucial element of risk assessment (FAO and WHO 2020). Uncertainty analysis (see above) is part of this quality assurance.

Following the uncertainty analysis, models should be both verified and validated. The verification can be achieved by making an audit of the model to ensure that it operates as intended by the developer. The audit can be internal or external with the publication of a draft document open to public consultation. Models should also be validated. The validation can be defined as a demonstration of the accuracy of the model. The validation of a risk model is usually done by comparing the model output with epidemiological data.

One of the current challenges in risk assessment is the difficulty of exploiting and applying the knowledge generated/present in a quantitative microbial risk assessment. This represents a severe obstacle for the establishment of timely risk assessment using the most up-to-date knowledge. Thus, the microbiological food safety community would benefit by having knowledge efficiently shared in a transparent and consistent way in order to facilitate its exploitation and application in microbiological risk assessment and management. Many of the developed models are published in non-standardized ways, which hinders efficient exchange, re-use and continuous improvement of models within the risk assessment domain (Filter *et al.* 2021).

This transparent and consistent way can be achieved, for instance, through the development of a user-friendly model repository that stores the knowledge in a standardized format

In this context, it is worth noting the creation of a standard-based, publicly available knowledge repository that is intended for this kind of knowledge sharing, RAKIP (https://foodrisklabs.bfr.bund.de/rakip-web-portal/). Specifically, this project aims at the development of harmonized data formats, controlled vocabularies and open-source software code libraries in order to facilitate knowledge transparency and consistency and exchange between software tools in QMRAs (Plaza-Rodríguez *et al.* 2018).

6.8. References

Augustin, J.-C., Kooh, P., Bayeux, T., Guillier, L., Meyer, T., Silva, J.-D., Villena, I., Sanaa, M., Cerf, O. (2020). Contribution of foods and poor food-handling practices to the burden of foodborne infectious diseases in France. *Foods*, 9(11), 1644.

Bartsch, C., Plaza-Rodriguez, C., Trojnar, E., Filter, M., Johne, R. (2019). Predictive models for thermal inactivation of human norovirus and surrogates in strawberry puree. *Food Control*, 96, 87–97.

Bergion, V., Lindhe, A., Sokolova, E., Rosén, L. (2018). Risk-based cost–benefit analysis for evaluating microbial risk mitigation in a drinking water system. *Water Research*, 132, 111–123.

Boué, G. (2017). Public health risk-benefit assessment in foods: Methodological development with application to infant milk-based diet. Thesis, Ecole nationale vétérinaire, Nantes.

Cassini, A., Colzani, E., Pini, A., Mangen, M.-J.J., Plass, D., McDonald, S.A., Maringhini, G., Van Lier, A., Haagsma, J.A., Havelaar, A.H. (2018). Impact of infectious diseases on population health using incidence-based disability-adjusted life years (DALYs): Results from the Burden of Communicable Diseases in Europe Study, European Union and European Economic Area Countries, 2009 to 2013. *Eurosurveillance*, 23(16), 17–00454.

Charpentier, A. and Cherrier, B. (2019). La valeur de la vie humaine. *Risques*, 118, 107–111.

Chen, Y., Dennis, S.B., Hartnett, E., Paoli, G., Pouillot, R., Ruthman, T., Wilson, M. (2013). FDA-iRISK – A comparative risk assessment system for evaluating and ranking food-hazard pairs: Case studies on microbial hazards. *Journal of Food Protection*, 76(3), 376–385.

Codex Alimentarius Commission (1999). Principles and guidelines for the conduct of microbiological risk assessment CAC/GL-30. Document, Codex Alimentarius International Food Standards.

Daniel, N., Casadevall, N., Sun, P., Sugden, D., Aldin, V. (2020). *The Burden of Foodborne Disease in the UK 2018*. Food Standards Agency, London.

De Oliveira Mota, J., Boué, G., Guillou, S., Pierre, F., Membré, J.-M. (2019). Estimation of the burden of disease attributable to red meat consumption in France: Influence on colorectal cancer and cardiovascular diseases. *Food and Chemical Toxicology*, 130, 174–186.

De Oliveira Mota, J., Guillou, S., Pierre, F., Membré, J.-M. (2020). Quantitative assessment of microbiological risks due to red meat consumption in France. *Microbial Risk Analysis*, 15, 100103.

Duret, S., Guillier, L., Hoang, H.-M., Flick, D., Laguerre, O. (2014). Identification of the significant factors in food safety using global sensitivity analysis and the accept-and-reject algorithm: Application to the cold chain of ham. *International Journal of Food Microbiology*, 180, 39–48.

Duret, S., Hoang, H.M., Derens-Bertheau, E., Delahaye, A., Laguerre, O., Guillier, L. (2019). Combining quantitative risk assessment of human health, food waste, and energy consumption: The next step in the development of the food cold chain? *Risk Analysis*, 39(4), 906–925.

EFSA (2007). Opinion of the scientific panel on biological hazards (BIOHAZ) on microbiological criteria and targets based on risk analysis. *EFSA Journal*, 5(3), 462.

EFSA Panel on Biological Hazards (2015). Scientific opinion on the development of a risk ranking toolbox for the EFSA BIOHAZ Panel. *EFSA Journal*, 13(1), 3939.

EFSA Panel on Biological Hazards, Ricci, A., Allende, A., Bolton, D., Chemaly, M., Davies, R., Fernández Escámez, P.S., Girones, R., Herman, L., Koutsoumanis, K. (2018). *Listeria monocytogenes* contamination of ready-to-eat foods and the risk for human health in the EU. *EFSA Journal*, 16(1), e05134.

FAO and WHO (2004). Report of the twentieth session of the Codex Committee on General Principles, 3–7 May, Paris, France. Report, FAO and WHO.

FAO and WHO (2020). Guidance of microbiological risk assessment for food. Joint FAO/WHO expert meetings on microbiological risk assessment (JEMRA) on methodologies of microbiological risk assessment. Report, FAO and WHO.

Filter, M., Sundermann, E.M., Mesa-Varona, O., Buschhardt, T., Lopez De Abechuco, E., Georgiadis, M. (2021). Minimum information required to annotate food safety risk assessment models (MIRARAM). *Food Research International*, 139, 109952.

Flage, R. and Aven, T. (2015). Emerging risk–conceptual definition and a relation to black swan type of events. *Reliability Engineering and System Safety*, 144, 61–67.

Garre, A., Boué, G., Fernández, P.S., Membré, J.-M., Egea, J.A. (2020). Evaluation of multicriteria decision analysis algorithms in food safety: A case study on emerging zoonoses prioritization. *Risk Analysis*, 40(2), 336–351.

Gkogka, E., Reij, M.W., Gorris, L.G.M., Zwietering, M.H. (2013a). The application of the appropriate level of protection (ALOP) and food safety objective (FSO) concepts in food safety management, using *Listeria monocytogenes* in deli meats as a case study. *Food Control*, 29(2), 382–393.

Gkogka, E., Reij, M.W., Gorris, L.G.M., Zwietering, M.H. (2013b). Risk assessment strategies as a tool in the application of the appropriate level of protection (ALOP) and food safety objective (FSO) by risk managers. *International Journal of Food Microbiology*, 167(1), 8–28.

Gkogka, E., Reij, M.W., Gorris, L.G.M., Zwietering, M.H. (2020). Risk assessment of *Clostridium perfringens* in Cornish pasties in the UK. *Food Control*, 108, 106822.

González, S.C., Possas, A., Carrasco, E., Valero, A., Bolívar, A., Posada-Izquierdo, G.D., García-Gimeno, R.M., Zurera, G., Pérez-Rodríguez, F. (2019). "MicroHibro": A software tool for predictive microbiology and microbial risk assessment in foods. *International Journal of Food Microbiology*, 290, 226–236.

Gorris, L.G. (2005). Food safety objective: An integral part of food chain management. *Food Control*, 16(9), 801–809.

Guillier, L., Kabunda, J.-M., Denis, J.-B., Albert, I. (2013). Elicitation for food microbial risk assessment: A probabilistic approach extending risk ranger proposal. *Journal de la Société française de statistique*, 154(3), 113–123.

Guillier, L., Duret, S., Hoang, H.-M., Flick, D., Nguyen-Thé, C., Laguerre, O. (2016). Linking food waste prevention, energy consumption and microbial food safety: The next challenge of food policy? *Current Opinion in Food Science*, 12, 30–35.

Haberbeck, L.U., Plaza-Rodríguez, C., Desvignes, V., Dalgaard, P., Sanaa, M., Guillier, L., Nauta, M., Filter, M. (2018). Harmonized terms, concepts and metadata for microbiological risk assessment models: The basis for knowledge integration and exchange. *Microbial Risk Analysis*, 10, 3–12.

Havelaar, A.H., Nauta, M.J., Jansen, J.T. (2004). Fine-tuning food safety objectives and risk assessment. *International Journal of Food Microbiology*, 93(1), 11–29.

Havelaar, A.H., Mangen, M.J.J., De Koeijer, A.A., Bogaardt, M.J., Evers, E.G., Jacobs-Reitsma, W.F., Van Pelt, W., Wagenaar, J.A., De Wit, G.A., Van Der Zee, H. (2007). Effectiveness and efficiency of controlling *Campylobacter* on broiler chicken meat. *Risk Analysis: An International Journal*, 27(4), 831–844.

ICMSF (2002). *Microorganisms in Foods 7: Microbiological Testing in Food Safety Management*. Springer, New York.

Karch, H., Denamur, E., Dobrindt, U., Finlay, B.B., Hengge, R., Johannes, L., Ron, E.Z., Tønjum, T., Sansonetti, P.J., Vicente, M. (2012). The enemy within us: Lessons from the 2011 European *Escherichia coli* O104:H4 outbreak. *EMBO Molecular Medicine*, 4(9), 841–848.

Koch, L., Lopes, A.-A., Maiguy, A., Guillier, S., Guillier, L., Tournier, J.-N., Biot, F. (2020). Natural outbreaks and bioterrorism: How to deal with the two sides of the same coin? *Journal of Global Health*, 10(2), 020317.

Lindhjem, H., Navrud, S., Braathen, N.A., Biausque, V. (2011). Valuing mortality risk reductions from environmental, transport, and health policies: A global meta-analysis of stated preference studies. *Risk Analysis*, 31(9), 1381–1407.

Mangen, M.-J.J., Plass, D., Havelaar, A.H., Gibbons, C.L., Cassini, A., Mühlberger, N., Van Lier, A., Haagsma, J.A., Brooke, R.J., Lai, T., De Waure, C., Kramarz, P., Kretzschmar, M.E.E., BCoDE Consortium (2013). The pathogen- and incidence-based DALY approach: An appropriated methodology for estimating the burden of infectious diseases. *PLoS One*, 8(11), e79740.

Membré, J.-M. and Boué, G. (2018). Quantitative microbiological risk assessment in food industry: Theory and practical application. *Food Research International*, 106, 1132–1139.

Membré, J.-M., Farakos, S.S., Nauta, M. (2020). Risk-benefit analysis in food safety and nutrition. *Current Opinion in Food Science*, 39, 76–82.

Mughini-Gras, L., Kooh, P., Fravalo, P., Augustin, J.-C., Guillier, L., David, J., Thébault, A., Carlin, F., Leclercq, A., Jourdan-Da-Silva, N., Pavio, N., Villena, I., Sanaa, M., Watier, L. (2019). Critical orientation in the jungle of currently available methods and types of data for source attribution of foodborne diseases. *Frontiers in Microbiology*, 10, 2578.

Nauta, M.J. (2000). Separation of uncertainty and variability in quantitative microbial risk assessment models. *International Journal of Food Microbiology*, 57(1–2), 9–18.

Plaza-Rodríguez, C., Haberbeck, L.U., Desvignes, V., Dalgaard, P., Sanaa, M., Nauta, M., Filter, M., Guillier, L. (2018). Towards transparent and consistent exchange of knowledge for improved microbiological food safety. *Current Opinion in Food Science*, 19, 129–137.

Pouillot, R. and Delignette-Muller, M.L. (2010). Evaluating variability and uncertainty separately in microbial quantitative risk assessment using two R packages. *International Journal of Food Microbiology*, 142(3), 330–340.

Pouillot, R. and Guillier, L. (2020). Understanding uncertainty and variability in risk assessment. In *Risk Assessment Methods for Biological and Chemical Hazards in Food*, Pérez-Rodríguez, F. (ed.). CRC Press, Boca Raton.

Prache, S., Sante-Lhoutellier, V., Adamiec, C., Astruc, T., Baéza, E., Bouillot, P.-E., Bugeon, J., Cardinal, M., Cassar-Malek, I., Clinquart, A. (2020). La qualité des aliments d'origine animale selon les conditions de production et de transformation. Thesis, INRAE, Paris.

Rieu, E., Duhem, K., Vindel, E., Sanaa, M. (2007). Food safety objectives should integrate the variability of the concentration of pathogen. *Risk Analysis*, 27(2), 373–386.

Sanaa, M. (2020). Dose–response models for microbial risk assessment. In *Risk Assessment Methods for Biological and Chemical Hazards in Food*, Pérez-Rodríguez, F. (ed.). CRC Press, Boca Raton.

Sassi, F. (2006). Calculating QALYs, comparing QALY and DALY calculations. *Health Policy and Planning*, 21(5), 402–408.

Strachan, N.J., Doyle, M.P., Kasuga, F., Rotariu, O., Ogden, I.D. (2005). Dose response modelling of *Escherichia coli* O157 incorporating data from foodborne and environmental outbreaks. *International Journal of Food Microbiology*, 103(1), 35–47.

Suijkerbuijk, A.W., Over, E.A., Opsteegh, M., Deng, H., Gils, P.F.V., Bonačić Marinović, A.A., Lambooij, M., Polder, J.J., Feenstra, T.L., Giessen, J.W.V.D. (2019). A social cost–benefit analysis of two one health interventions to prevent toxoplasmosis. *PloS One*, 14(5), e0216615.

Téhard, B., Detournay, B., Borget, I., Roze, S., De Pouvourville, G. (2020). Value of a QALY for France: A new approach to propose acceptable reference values. *Value in Health*, 23(8), 985–993.

Tenenhaus-Aziza, F. and Ellouze, M. (2015). Software for predictive microbiology and risk assessment: A description and comparison of tools presented at the ICPMF8 Software Fair. *Food Microbiology*, 45, 290–299.

Tuominen, P., Ranta, J., Maijala, R. (2007). Studying the effects of POs and MCs on the salmonella ALOP with a quantitative risk assessment model for beef production. *International Journal of Food Microbiology*, 118(1), 35–51.

Van Schothorst, M. (2002). Implementing the results of a microbiological risk assessment: Pathogens risk management. In *Microbiological Risk Assessment in Food Processing*, Brown, M. and Stringer, M. (eds). CRC Woodhead Publishing, Cambridge.

Van Schothorst, M. (2005). A proposed framework for the use of FSOs. *Food Control*, 16(9), 811–816.

Van Wagenberg, C., Van Horne, P., Sommer, H.M., Nauta, M. (2016). Cost-effectiveness of *Campylobacter* interventions on broiler farms in six European countries. *Microbial Risk Analysis*, 2, 53–62.

Conclusion

Jeanne-Marie MEMBRÉ
SECALIM, INRAE, Oniris, Nantes, France

The purpose of microbiological risk assessment is to characterize the nature and probability of the damage resulting from human exposure to the biological agents present in foodstuffs. Ever since 1999, the World Health Organization (WHO) and the Food and Agriculture Organization of the United Nations (FAO), through the Codex Alimentarius Commission (CAC), have been a reference on this topic, both by establishing general principles and publishing practical guidelines for assessing the microbiological risks in food. This book was therefore inspired by their work, particularly in its structure. The first chapter of this book dealt with the first stage of microbiological risk assessment, namely hazard identification; the second and third chapters dealt with exposure assessment; the fourth and fifth with hazard characterization; and finally the sixth with risk characterization.

In the context of the production, distribution and consumption of food products, the risks must take into account all stages of the chain: from the production of raw materials ("farm") to consumption ("fork") and even through to ingestion ("human"). Microbiological risk assessment thus covers the "farm-to-fork-to-human" continuum. Of course, depending on the food product, its composition, its manufacture and its packaging, the key steps to be taken into account in the risk assessment may be mostly upstream or mostly downstream of this continuum.

An analysis of the scientific output on the theme of microbiological risk assessment associated with the food processing/distribution chain is presented below. It shows that this research topic is dynamic. A focus on the most recent years (2015–2020) also makes it possible to confirm the interest in the hazards associated with the most frequent causes of foodborne illness and death, namely noroviruses, *Campylobacter* spp. and non-typhoid *Salmonella* spp. The *Escherichia coli* pathogens are also an important topic for investigation.

This literature analysis is followed by a presentation of possible avenues for the development of the theme, or more precisely its integration into a more general approach through successive inclusion: microbiological risks □ ⊂ health risks ⊂ public health risks ⊂ public health risks and environmental risks, etc.

Scientific work on microbiological risk assessment

A Web of Science search (January 2021) using the query "(((TITLE: (risk) AND TITLE: (evaluation OR assessment)) AND TOPIC: (Food)) AND TOPIC: (microbiol* OR microorgan))" brought up more than 1,037 scientific articles.

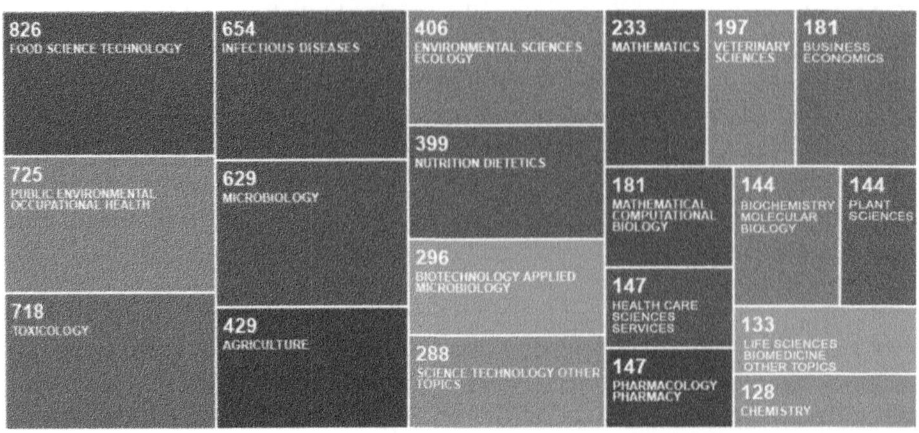

Figure C.1. *Breakdown of the 1,037 published articles on microbiological risk assessment by major scientific theme. For a color version of this figure, see www.iste.co.uk/membre/microbiological.zip*

Most of these articles are categorized under the topic of Food Science Technology (Figure C.1) and they are mainly published in two scientific journals: the *International Journal of Food Microbiology* (107 articles) and the *Journal of Food Protection* (101 articles).

A targeted analysis of the 294 articles from among the 1,037 initial articles that were cited at least twice each year on average shows that scientific activity around microbiological risk assessment really took off at the end of the 1990s and in the early 2000s, with benchmark articles being published. As an example, the article by Cassin *et al.* (1998) on *E. coli* O157:H7 on meat and the one by Rosenquist *et al.* (2003) on *Campylobacter* on chicken have been cited, respectively, 13 and 18 times on average per year since their publication.

A focus on the most recent years (2015–2020) is one way to identify the most current topics of interest for researchers and even emerging trends or topics. There were 372 articles published between 2015 and 2020. We can now confirm the interest in the hazards associated with the most frequent causes of foodborne illness and death, namely noroviruses, *Campylobacter* spp and non-typhoid *Salmonella* spp. The *Escherichia coli* pathogens are currently also a major topic for investigation.

Among the emerging topics or topics of growing interest, it is worth noting water and its (re)use for drinking purposes or in the food industry (Table C.1), raw milk (Table C.2) and resistance to antibiotics (Table C.3). A final topic of current interest that could also take on more importance is consumer practices and/or changes in practices (Table C.4).

Title	DOI
Development of a *Cryptosporidium*-arsenic multi-risk assessment model for infant formula prepared with tap water in France	10.1016/j.foodres.2018.03.054
Zero risk does not exist: Lessons learned from microbial risk assessment related to use of water and safety of fresh produce	10.1111/1541-4337.12140
Microbial risk assessment of drinking water based on hydrodynamic modelling of pathogen concentrations in source water	10.1016/j.scitotenv.2015.04.040
A farm to fork risk assessment for the use of wastewater in agriculture in Accra, Ghana	10.1371/journal.pone.0142346

Title	DOI
Quantitative microbial risk assessment of Greywater on-site reuse	10.1016/j.scitotenv.2018.04.197
Quality of greywater treated in biochar filter and risk assessment of gastroenteritis due to household exposure during maintenance and irrigation	10.1111/jam.13273
Probabilistic quantitative microbial risk assessment model of norovirus from wastewater irrigated vegetables in Ghana using genome copies and fecal indicator ratio conversion for estimating exposure dose	10.1016/j.scitotenv.2017.05.168
Epidemiological study for the assessment of health risks associated with graywater reuse for irrigation in arid regions	10.1016/j.scitotenv.2015.08.009
Quantitative microbial risk assessment of Cryptosporidium and Giardia in well water from a native community of Mexico	10.1080/09603123.2014.989492
Wastewater treatment and public health in Nunavut: a microbial risk assessment framework for the Canadian Arctic	10.1007/s11356-017-8566-8
Quantitative microbial risk assessment and opportunist waterborne infections – Are there too many gaps to fill?	10.3390/ijerph15061150
Quantitative microbial risk assessment and infectious disease transmission modeling of waterborne enteric pathogens	10.1007/s40572-018-0196-x
Evaluation of primitive ground water supplies as a risk factor for the development of major waterborne zoonosis in Egyptian children living in rural areas	10.1016/j.jiph.2017.07.025
Public health and economic risk assessment of waterborne contaminants and pathogens in Finland	10.1016/j.scitotenv.2017.05.007
Implementation of quantitative microbial risk assessment (QMRA) for public drinking water supplies: Systematic review	10.1016/j.watres.2020.115614
Risk-based assessment and criteria specification of the microbial safety of wastewater reuse in food processing: Managing *Listeria monocytogenes* contamination in pasteurized fluid milk	10.1016/j.watres.2020.115466
Validation of quantitative microbial risk assessment using epidemiological data from outbreaks of waterborne gastrointestinal disease	10.1111/risa.13189

Title	DOI
Effect of unboiled water consumption data on sensitivity analysis in quantitative microbial risk assessment	10.1038/s41545-018-0018-6
A probabilistic assessment of the contribution of wastewater-irrigated lettuce to *Escherichia coli* O157:H7 infection risk and disease burden in Kumasi, Ghana	10.2166/wh.2014.108

Table C.1. *Recent articles (titles and DOIs) on the topic of water and its use for drinking purposes or its reuse in the food industry*

Title	DOI
Quantitative risk assessment of human salmonellosis and listeriosis related to the consumption of raw milk in Italy	10.4315/0362-028X.JFP-14-171
Consumers' behavior in quantitative microbial risk assessment for pathogens in raw milk: Incorporation of the likelihood of consumption as a function of storage time and temperature	10.3168/jds.2015-10175
Assessment of raw milk quality and stakeholders' awareness on milk-borne health risks in Arusha City and Meru District, Tanzania	10.1007/s11250-015-0810-y
Paediatric HUS cases related to the consumption of raw milk sold by vending machines in Italy: Quantitative risk assessment based on *Escherichia coli* O157 official controls over 7 years	10.1111/zph.12331

Table C.2. *Recent articles (titles and DOIs) on the topic of raw milk*

Title	DOI
Unraveling the role of vegetables in spreading antimicrobial-resistant bacteria: A need for quantitative risk assessment	10.1089/fpd.2018.2501
Risk assessments evaluating foodborne antimicrobial resistance in humans: A scoping review	10.1016/j.mran.2018.08.002

Quantitative risk assessment of antimicrobial-resistant foodborne infections in humans due to recombinant bovine somatotropin usage in dairy cows	10.4315/0362-028X.JFP-16-404
Microbial risk assessment and antimicrobial resistance	10.1016/B978-0-12-818882-8.00020-6

Table C.3. *Recent articles (titles and DOIs) on the topic of antibiotic resistance*

Title	DOI
Food consumption and handling survey for quantitative microbiological consumer phase risk assessments	10.4315/0362-028X.JFP-15-448
Modelling and magnitude estimation of cross-contamination in the kitchen for quantitative microbiological risk assessment (QMRA)	10.2903/j.efsa.2020.e181106
An assessment of listeriosis risk associated with a contaminated production lot of frozen vegetables consumed under alternative consumer handling scenarios	10.4315/0362-028X.JFP-19-092
Development of a risk-ranking framework to evaluate simultaneously biological and chemical hazards related to food safety: Application to emerging dietary practices in France	10.1016/j.foodcont.2020.107279

Table C.4. *Recent articles (titles and DOIs) on the topic of the consumer and consumer practices*

Integration of microbiological risks into a more general risk assessment process

As shown in Figure C.2, microbiological risks are only one aspect of the risks to be considered in a general risk assessment process.

Microbiological risks are part of food safety risks (level 2 in Figure C.2), alongside chemical risks; their assessment and control contributes to food safety. For a given food or eating practice, simultaneous evaluation is still difficult. Risk ranking-type approaches are necessary (Van der Fels-Klerx *et al.* 2018; FAO 2020). In the future, they will require methodological developments in the aggregation of risks, such as the methodology applied recently to new consumer practices (Eygue *et al.* 2020).

At level 3, we have the public health risks (Figure C.2). In fact, food safety risks themselves are only one part of public health risks: nutritional risks must not be overlooked. For example, it has been estimated that in France nutritional risks such as colorectal cancer and cardiovascular diseases associated with the consumption of red meat are greater than microbiological risks due to the bacteria and viruses present in the meat (De Oliveira Mota *et al.* 2021). This risk comparison is part of the new scientific discipline called "health benefit risk assessment" (Membré *et al.* 2021).

Finally, in a sustainability approach, health risks must be put into perspective with environmental risks. The quantitative assessment of health risks and nutritional risks is then integrated into life-cycle analyses (Guinée *et al.* 2017; Dong *et al.* 2018; Hollander *et al.* 2019; Kesse-Guyot *et al.* 2020). Without a doubt, this fourth level of integration (Figure C.2) should be pursued in terms of both methodology and application. It is actually an integration of public health risks and environmental risks that enables the deployment of healthy and sustainable food systems.

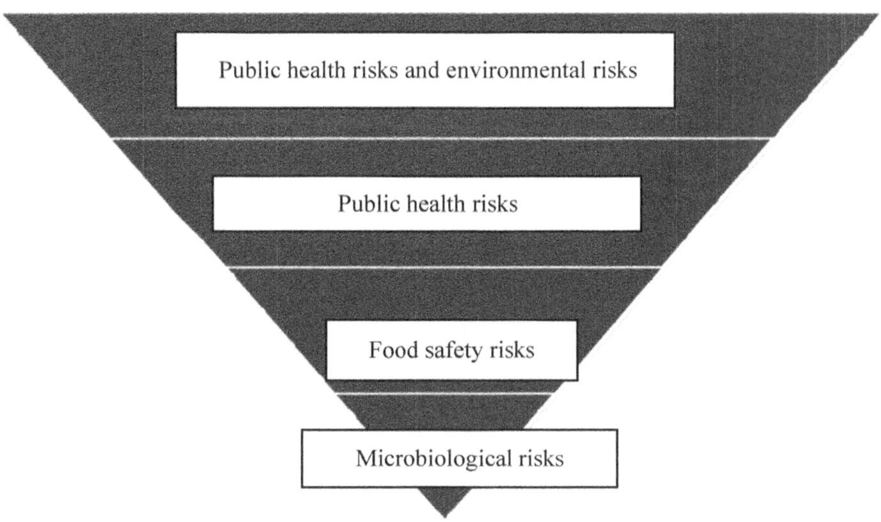

Figure C.2. *Microbiological risks repositioned in a more general risk assessment process*

Let us now return in more detail to risk ranking, as recently proposed by the FAO (FAO 2020).

Risk ranking of hazards

Food safety hazard ranking is the systematic analysis and categorization of food and/or food hazards in terms of public health risks based on the probability and severity of negative impacts on human health in a target population. Risk ranking of hazards provides national food safety authorities with a scientific basis for making informed regulatory decisions, improving disease surveillance, determining priorities for food inspections and educating the public about threats to food safety, etc. (FAO 2020).

Categorizing foodborne hazards in terms of risks to public health, based on the probability and severity of negative impacts on human health, means to some extent aggregating the results of several risk assessments.

By way of example, the risk ranking carried out by the FAO in its case study on microbiological risks is shown in the graph below. Risks are expressed in disability-adjusted life years (DALYs) (Murray and Acharya 1997), a unit used by the WHO in its work on disease burden (WHO 2018). It should be noted that this example is given only to illustrate the method proposed by the FAO; it should not be interpreted in absolute values.

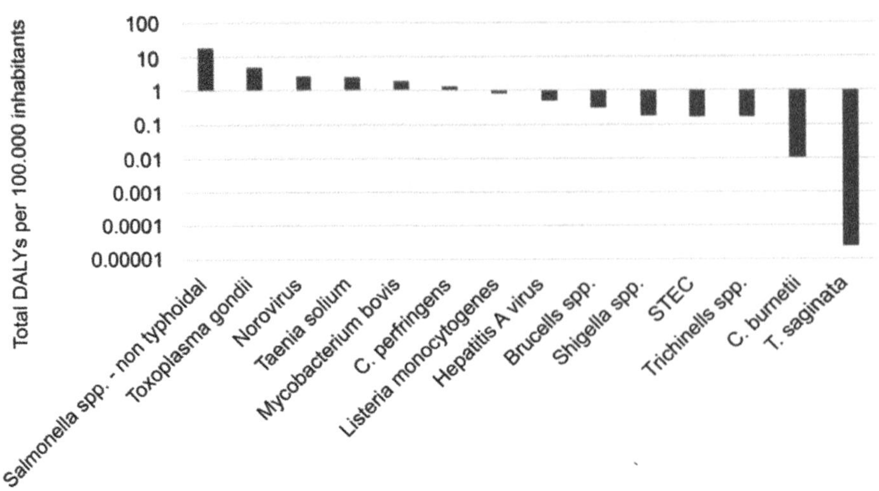

Figure C.3. *Example of categorization of microbiological risks based on their probability and severity. Example adapted from the FAO (2020)*

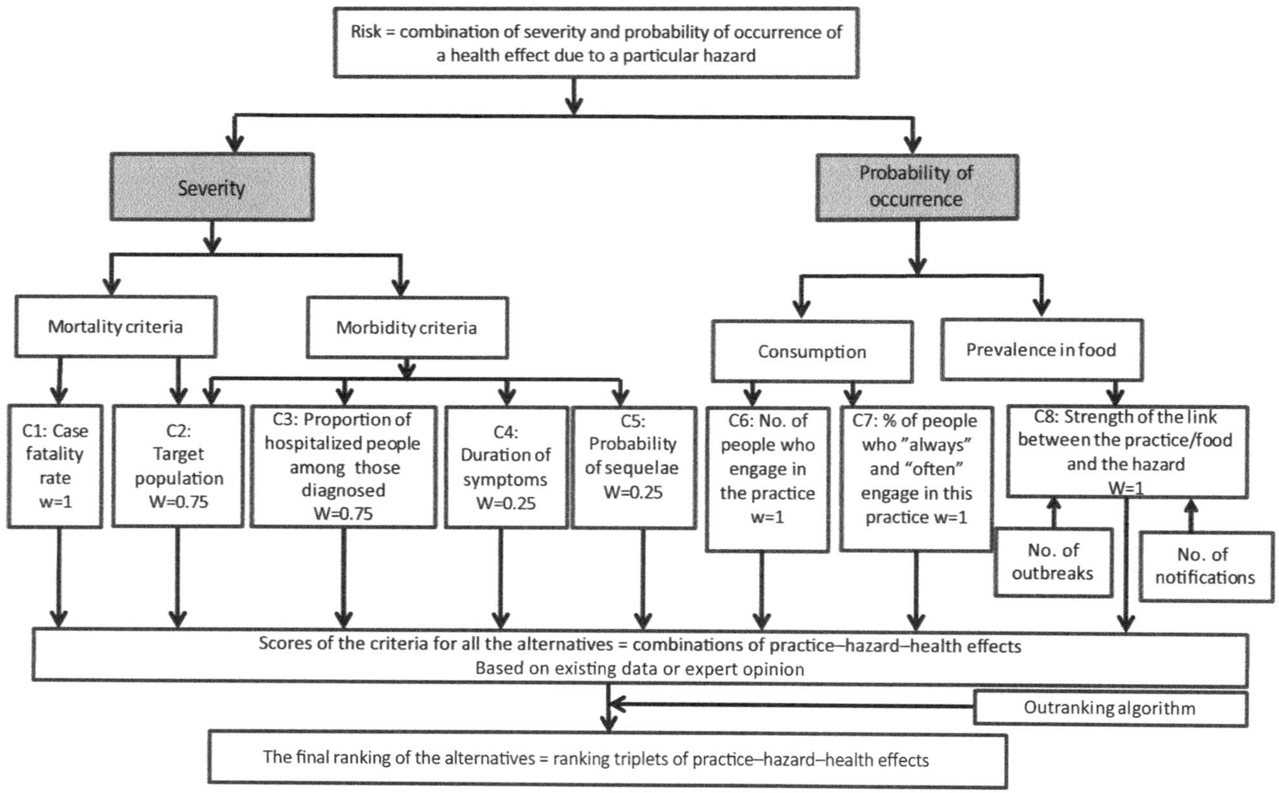

Figure C.4. *Method for simultaneous risk ranking of microbiological and chemical hazards proposed by Eygue* et al. *(2020) in their work on new consumer practices. For a color version of this figure, see www.iste.co.uk/membre/microbiological.zip*

Other examples of methods dedicated to the prioritization risk ranking of hazards have been put forward in the literature; in particular, there is one method developed by ANSES which categorizes food–hazard pairings, on the one hand with microbiological criteria and on the other hand with chemical criteria (ANSES 2020).

As mentioned above, simultaneous microbiological and chemical risk assessment methods are still under construction. Eygue *et al.* (2020) used as a basis the work of the FAO (2017) to put forward a simultaneous risk ranking of chemical and microbiological risks based on eight criteria: five for severity and three for probability. The method has been applied to the new practices of French consumers. For each practice, a list of hazards was established, and for each hazard, a list of adverse health effects was established. Thus, ultimately, a rating was given to a practice–hazard–health effect triplet. This process is illustrated in Figure C.4.

References

ANSES (2020). Méthodologie de hiérarchisation des dangers biologiques et chimiques dans les aliments [Online]. Available at: https://www.anses.fr/fr/system/files/BIORISK2016SA0153Ra.pdf.

Cassin, M.H., Lammerding, A.M., Todd, E.C.D., Ross, W., McColl, R.S. (1998). Quantitative risk assessment for *Escherichia coli* O157:H7 in ground beef hamburgers. *International Journal of Food Microbiology*, 41(1), 21–44.

De Oliveira Mota, J.D.O., Guillou, S., Pierre, F., Membré, J.-M. (2021). Public health risk-benefit assessment of red meat in France: Current consumption and alternative scenarios. *Food and Chemical Toxicology*, 149, 111994.

Dong, Y., Miraglia, S., Manzo, S., Georgiadis, S., Sørup, H.J.D., Boriani, E., Hald, T., Thöns, S., Hauschild, M.Z. (2018). Environmental sustainable decision making – The need and obstacles for integration of LCA into decision analysis. *Environmental Science and Policy*, 87, 33–44.

Eygue, M., Richard-Forget, F., Cappelier, J.-M., Pinson-Gadais, L., Membré, J.-M. (2020). Development of a risk-ranking framework to evaluate simultaneously biological and chemical hazards related to food safety: Application to emerging dietary practices in France. *Food Control*, 115, 107279.

FAO (2017). Food safety risk management evidence-informed policies and decisions, considering multiple factors. Report, FAO Guidance Materials, Rome [Online]. Available at: http://www.fao.org/3/i8240en/I8240EN.pdf [Accessed 18 March 2020].

FAO (2020). FAO guide to ranking food safety risks at the national level. Report, FAO, Rome.

Guinée, J.B., Heijungs, R., Vijver, M.G., Peijnenburg, W. (2017). Setting the stage for debating the roles of risk assessment and life-cycle assessment of engineered nanomaterials. *Nature Nanotechnology*, 12(8), 727–733.

Hollander, A., De Jonge, R., Biesbroek, S., Hoekstra, J., Zijp, M.C. (2019). Exploring solutions for healthy, safe, and sustainable fatty acids (EPA and DHA) consumption in the Netherlands. *Sustainability Science*, 14(2), 303–313.

Kesse-Guyot, E., Chaltiel, D., Wang, J., Pointereau, P., Langevin, B., Allès, B., Rebouillat, P., Lairon, D., Vidal, R., Mariotti, F., Egnell, M., Touvier, M., Julia, C., Baudry, J., Hercberg, S. (2020). Sustainability analysis of French dietary guidelines using multiple criteria. *Nature Sustainability*, 3(5), 377–385.

Membré, J.-M., Santillana Farakos, S., Nauta, M. (2021). Risk–benefit analysis in food safety and nutrition. *Current Opinion in Food Science*, 39, 76–82.

Murray, C.J. and Acharya, A.K. (1997). Understanding DALYs (disability-adjusted life years). *Journal of Health Economics*, 16(6), 703–730.

Rosenquist, H., Nielsen, N.L., Sommer, H.M., Norrung, B., Christensen, B.B. (2003). Quantitative risk assessment of human campylobacteriosis associated with thermophilic *Campylobacter* species in chickens. *International Journal of Food Microbiology*, 83(1), 87–103.

Van Der Fels-Klerx, H.J., Van Asselt, E.D., Raley, M., Poulsen, M., Korsgaard, H., Bredsdorff, L., Nauta, M., D'Agostino, M., Coles, D., Marvin, H.J.P., Frewer, L.J. (2018). Critical review of methods for risk ranking of food-related hazards, based on risks for human health. *Critical Reviews in Food Science and Nutrition*, 58(2), 178–193.

WHO (2018). Health statistics and information systems: About the global burden of disease (GBD) project [Online]. Available at: http://www.who.int/healthinfo/global_burden_disease/about/en/ [Accessed 5 April 2018].

Appendices

Jeanne-Marie MEMBRÉ
SECALIM, INRAE, Oniris, Nantes, France

Appendix 1. Hazard identification – Example of cooked ham

This example concerns boned, formed cooked ham. It is taken from an article by Zwietering *et al.* (2016).

At manufacture, the pork joints are cut and boned; after the stages of brining, tumbling and forming in a plastic case, the product is thermally processed (Figure A.1). During the thermal processing, the internal cooking temperature (core temperature) is above 70 °C and the target pasteurization value is more than 40 minutes. This thermal process is sufficient to completely inactivate the *Listeria monocytogenes* cells that can potentially be brought in with the raw materials or introduced by the stages of the operational process prior to cooking (Carlier *et al.* 1996).

After thermal processing, the ham is sliced, packaged and then stored under refrigerated conditions (Figure A.1). During mechanical slicing, recontamination of the cooked ham by *L. monocytogenes* may occur (Keskinen *et al.* 2008; Sheen and Hwang 2008). Sources of recontamination could be another piece of contaminated product (cross-contamination) or a biofilm that has survived the cleaning and disinfection procedure. The level of recontamination was estimated by analyzing various bibliographic sources (Table A.1). On average, 3.2% of ham slices are contaminated at the time when they are distributed to retail stores. These figures are in line with surveys carried out by EU Member States (EFSA 2014): on average, 5.1% of samples of pork in cooked, ready-to-eat products sampled in retail stores in 2011 and 2012 tested positive for *L. monocytogenes* (median = 3.5%, range 0 to 26.3%). Clearly, an average value of 3.2% means that some batches of cooked ham have a much lower prevalence (some may even not be contaminated at all) and other batches have a higher prevalence. It is important to bear in mind that the analysis of products collected at retail outlets generates uncertainty, even bias, due to the potential growth of *L. monocytogenes* and differences between the unit weights of the samples and products.

Products and origin	Sample size (g)	No. of positive samples/no. of samples	Prevalence (%)	Reference
Luncheon meats, USA	25	82/9199	0.89	Gombas *et al.* (2003)
Ham, Brazil	25	1/65	1.5	Martins Germano (2011)
Ham, United Kingdom	100	40/949	4.2	Little *et al.* (2009)
Cooked ham, Belgium	25	54/879	6.1	Uyttendaele *et al.* (1999)
Prevalence estimates			Mean 3.2	

Table A.1. *Frequency of contamination of sliced ham by* L. monocytogenes, *adapted from Zwietering* et al. *(2016)*

Next, the ham slices are hermetically packaged until the time of their consumption. There are thus no additional sources of recontamination. However, *L. monocytogenes* is capable of multiplying on ham even when the product is stored under refrigerated conditions (Beumer *et al.* 1996; Mataragas *et al.* 2006; Hwang *et al.* 2012; Zhang *et al.* 2012) and it can then reach concentrations that are hazardous for the consumer.

A search in RASFF confirmed this literature analysis. The search was carried out with no date restrictions, using the keywords "Listeria" and "ham", or "Listeria" and "cooked". From the first 51 results, after eliminating duplicates and checking the basic text of the alert in order to focus as closely as possible on the product "cooked sliced ham", 20 cases were found. Since 2004, there have been regular consumer alerts for ham-based products

(Table A.2). The notifications were the result either of tests specific to each company or of official tests of products on the market.

The hazard identification carried out clearly gives a good reason for keeping *Listeria monocytogenes* on the primary list of hazards and even makes it possible to identify the stage that should be monitored in particular: the slicing of the ham after cooking and before packaging. The article concludes with the need to set up a control system for *L. monocytogenes* focusing on the slicers (environment, equipment in contact with the product, floors, etc.).

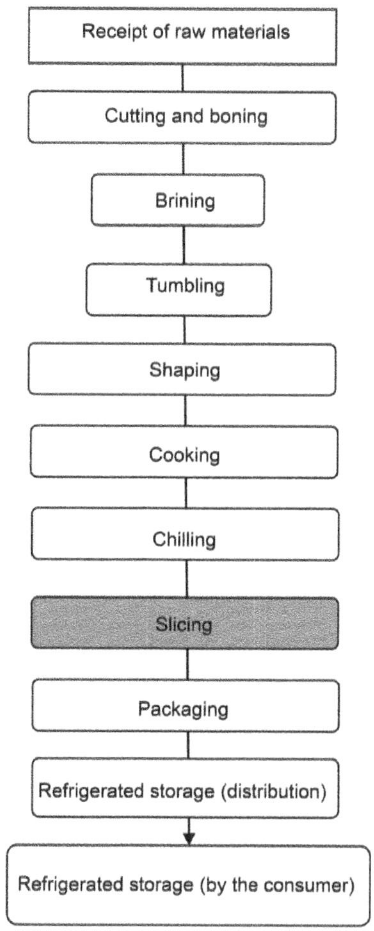

Figure A.1. *Diagram of cooked ham production, adapted from Zwietering* et al. *(2016)*

Year	Notification type	Notification basis	Subject
2004	Alert	Company's own check	*L. monocytogenes* in sliced ham
2004	Information	Official control on the market	*L. monocytogenes* (10^4 CFUs/g) in egg salad with ham
2005	Information	Official control on the market	*L. monocytogenes* (<50 CFUs/g) in ham from Cyprus
2007	Alert	Official control on the market	*L. monocytogenes* (>100 CFUs/g) in cooked ham from Italy
2008	Alert	Company's own check	*L. monocytogenes* in vacuum-packed ham from the Czech Republic
2009	Information	Official control on the market	*L. monocytogenes* (<10/25 g) in sliced cooked ham from Germany
2009	Information	Official control on the market	*L. monocytogenes* (presence/25 g) in ham from Ireland
2009	Information	Official control on the market	*L. monocytogenes* (presence/25 g) in sliced baked ham from Ireland
2009	Information	Official control on the market	*L. monocytogenes* (10 CFUs/g) in vacuum-packed ham from Poland
2010	Information	Company's own check	*L. monocytogenes* (2,000 CFUs/g) in cooked ham from Germany
2010	Information	Company's own check	*L. monocytogenes* in rindless cooked ham from France
2010	Alert	Official control on the market	*L. monocytogenes* (5 out of 5 samples/25 g) in ham from Germany
2011	Alert	Official control on the market	*L. monocytogenes* (4,800 CFUs/g) in cooked ham from Italy
2011	Alert	Company's own check	*L. monocytogenes* (presence/25 g) in ham from Italy
2012	Alert	Company's own check	*L. monocytogenes* (<40 CFUs/g) in chilled ham from Spain
2012	Alert	Company's own check	*L. monocytogenes* (presence/25 g) in ham from Belgium
2012	Alert	Official control on the market	*L. monocytogenes* in spiced and cooked pork flank from Germany
2013	Alert	Official control on the market	*L. monocytogenes* in frozen cooked sliced meat product from Poland
2014	Alert	Company's own check	*L. monocytogenes* (presence/25 g) in ham from Germany

Table A.2. *Notifications for* L. monocytogenes *in cooked ham (or a similar product). Data extracted from RASFF in September 2014*

Appendix 2. Hazard identification – Example of refrigerated vacuum-packed products

This example concerns refrigerated vacuum-packed products. It is taken from an article by Membré *et al.* (2009).

The European chilled food sector expanded by 50% between 1991 and 1996 to reach 9.1 billion euros. In 2005, the European market for chilled prepared foods grew by 10.6 billion euros (Del Torre *et al.* 2004). In order to ensure high quality in terms of taste and aroma, the food safety of refrigerated and processed foods of extended durability (REPFEDs) relies on low-temperature thermal processing and refrigerated storage, sometimes in combination with other barrier measures such as reducing pH and water activity. The shelf lives of REPFEDs are generally from two to six weeks, depending on the process and formulation.

In REPFEDs, non-proteolytic *Clostridium botulinum* is regarded as a serious hazard due to its ability to grow at temperatures as low as 3°C (Notermans *et al.* 1990; Peck 1997; Carlin *et al.* 2004). The neurotoxin produced by *C. botulinum* is one of the most toxic natural toxins known. The toxic dose of foodborne botulism is very low. Consuming a mere 30 ng of the neurotoxin is enough to cause illness and even death (Lund and Peck 2000).

Thermal processes affect the growth of the spores of non-proteolytic *C. botulinum* in two ways: through inactivation and through stress.

It is generally accepted that a temperature of 90°C for 10 minutes brings about inactivation of 6 log of non-proteolytic *C. botulinum* (Gould 1999). The z values (temperature increase required to reduce the decimal reduction time by one logarithm) used to calculate process equivalence differ depending on guidelines and good practice guides, but, for example, the US FDA references a z value of 7°C for temperatures below 90°C (U.S. Food and Drug Administration 2001).

In addition to the mortal effect, there may be damage to some of the spores that remain after thermal processing. Due to this damage, the recovery time of the spores is prolonged and the latency period is increased (Peck *et al.* 1995; Peck 1997). This effect of the thermal processing is also dependent on the conditions of recovery, such as temperature (refrigerated environment), pH and water activity (aw). If the shelf life of the REPFED is short enough, it is biologically possible for the damaged *C. botulinum* spores to have enough time to recover, germinate, multiply and potentially produce toxins.

The objective of Membré *et al.*'s study was therefore to explore the possibility of identifying thermal processes that are more milde than 10 minutes at 90°C, integrating this latency period with the classic thermal inactivation effect. An exposure assessment model based on the concept of degree of protection (DoP) was developed. The principle of the approach used to calculate the degree of protection, combining the thermal inactivation effect and the thermal injury effect, is based on work from similar studies previously reported in the literature (Hauschild 1982; Lund 1993; Schillinger *et al.* 1996). Details are provided in equations [A.1]–[A.5]

$$\text{DoP} = \Delta\text{R} + \text{ThI} \qquad \text{[A.1]}$$

where ΔR is the decimal reduction due to thermal inactivation and ThI is the degree of protection due to thermal stress. Both are expressed as the decimal logarithm of the reciprocal of a probability:

$$\Delta\text{R} = \log(1/\text{Pr}) \qquad \text{[A.2]}$$

$$\text{ThI} = \log(1/\text{Pi}) \qquad \text{[A.3]}$$

with Pr being the probability that a spore will survive thermal processing and Pi the probability that the latency period will be shorter than the shelf life (SL):

$$\text{Pi} = \text{Probability}(\text{Lag} \leq \text{SL}) \qquad \text{[A.4]}$$

The concept of the DoP can be compared to an isoprobability method, as the DoP is derived from probability calculations. For example, there is the same probability (1.10^{-6}) (Figure A.2):

– of having a stability of 50 days with a product with pH 5.7 thermally processed at 84°C for 10 minutes;

– of having a stability of 30 days with a product with pH 6.3 thermally processed at 88°C for 10 minutes or a product with pH 6.0 thermally processed at 87°C for 5 minutes;

– of having a stability of 20 days with a product with pH 6.0 thermally processed at 84°C for 7 minutes;

– of having a stability of 25 days with a product with pH 6.0 thermally processed at 85°C for 10 minutes.

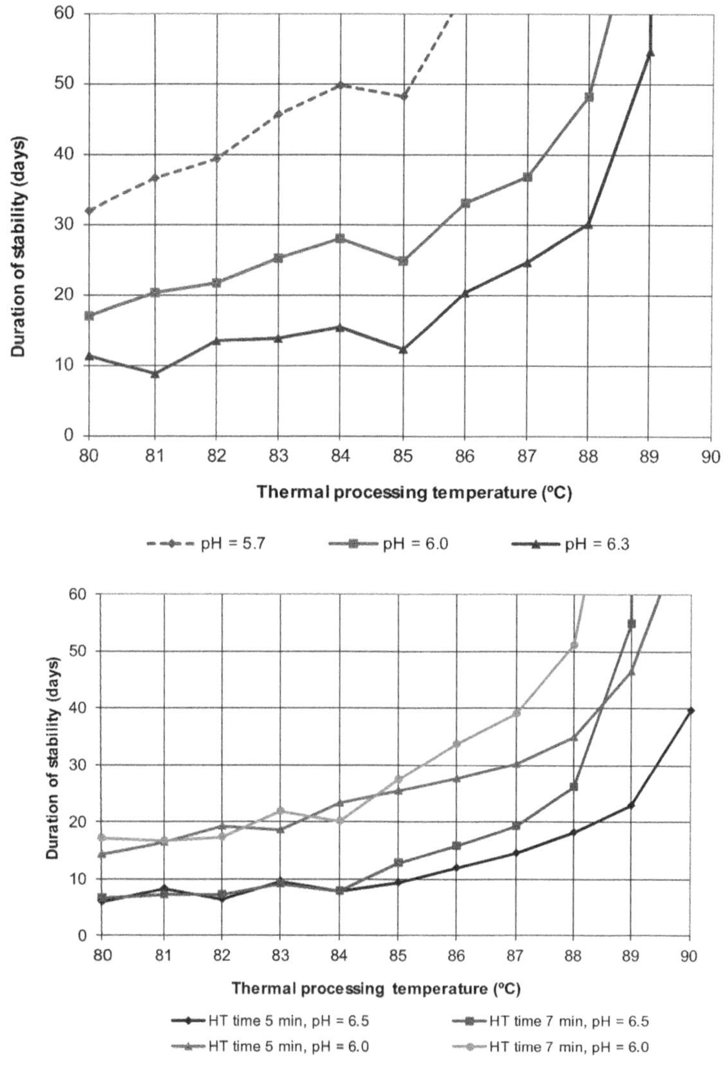

Figure A.2. *Combination of thermal processing and pH guaranteeing a 6 log degree of protection against non-proteolytic* C. botulinum *to provide stability in REPFEDs for a given period (y-axis). Top graph: processing time of 10 min; bottom graph: processing time of 5 or 7 min. For a color version of this figure, see www.iste.co.uk/membre/microbiological.zip*

These few combinations are only illustrations since the model enables the calculation of a whole range of thermal processing, pH and refrigerated storage time values permitting the same target to be achieved, that is, a degree of protection of 6 log.

The degree of protection approach illustrated here with *C. botulinum* and REPFEDs allows better use to be made of combined storage conditions, resulting in less damage to the quality of the product but nevertheless offering the required level of product safety.

The importance of these interactions has been recognized for the control of mesophilic spores in environmentally stable products. Braithwaite *et al.* (1971) produced a set of contours where F_0 limit values were plotted against aw and pH, clearly showing the combinations of conditions that require only relatively gentle thermal processing (e.g. F_0 0.3) to ensure ambient stability.

The idea of taking into account both destruction and stress was first introduced by Pivnick and Petrasovits (1973) in relation to the safety of canned salted meat with a long shelf life and then developed by Hauschild and Simonsen (1985) to estimate the "protection" of a food based on the probability that the spores that survive the thermal process will overcome inhibition, grow and produce a toxin.

While the outcomes described in this article are encouraging, the authors nevertheless conclude that there are limits. They stress the importance of validating such studies with different strains and different types of products from the REPFED category.

Finally, the authors conclude their article by noting that the degree of protection of 6 log is similar to a performance criterion of 6 log used in risk characterization to make the link between the operational scales applied throughout the production–processing–distribution chain of a product and consumer health.

Appendix 3. Source attribution

The following study was carried out as part of ANSES' BIORISK collective. It was published by Augustin *et al.* (2020).

Several hazard values were estimated, including incidence rate and disease severity (Table A.3). Hazard summaries are also available in the article. A summary of this information is provided below.

The incidences of foodborne diseases were drawn from a study carried out in France for the period 2008–2013 using surveillance data collected by the National Public Health Agency (Santé Publique France) and the National Reference Centers (CNR). Disease severity was expressed as disability-adjusted life years (DALYs), which take into account morbidity and mortality (Murray and Acharya 1997). Published studies (Havelaar *et al.* 2012; WHO 2017) were used to calculate the DALY scores for diseases. However, these values were sometimes modified by the ANSES group of experts. In these cases, the scores assigned were derived from those for diseases characterized by similar symptoms.

Hazard	DALYs in the Netherlands (Havelaar *et al.* 2012)	DALYs in Western Europe (WHO 2017)	Severity score (ANSES)	Incidence
Bacillus cereus	2.3	–	1	5
Campylobacter spp.	41	20	2	5
Clostridium botulinum	–	–	3	1
Listeria monocytogenes	1,450	8,000	4	2
Non-typhoid *Salmonella*	49	70	2	5
Shiga toxin-producing *E. coli*	143	20	2	4
Norovirus	2.4	2	1	5

Table A.3. *Foodborne disease severity scores and estimates of the foodborne morbidity burden in France (around 2010). More information on the hazards is listed in Table A.4 (Augustin* et al. *2020)*

Hazard	Short description
Bacillus cereus	*Bacillus cereus* outbreaks are mainly associated with the consumption of refrigerated processed foods of extended durability or mixed meals containing cereals. Outbreaks usually occur under improper storage conditions (disruption of the cold chain or shelf life being exceeded) or during preparation when conditions are not properly monitored (chilling and cooking) for homemade meals.

Campylobacter spp.	The main source of the foodborne campylobacteriosis considered was poultry meat. International attribution studies also link *Campylobacter* infections to poultry. Although many foods can act as a vector for *Campylobacter*, given that cross-contamination can occur during food preparation, poultry meat was considered only as the ultimate source of contamination. The factors promoting the appearance of campylobacteriosis are cross-contamination and undercooking of poultry meat.
Clostridium botulinum	Botulism is usually associated with the consumption of homemade foods, such as badly prepared, badly stored or badly cooked pork or canned products. A few cases of infant botulism have been attributed to the consumption of honey.
Listeria monocytogenes	Cases of listeriosis are linked to the consumption of a wide variety of foods. Most often, the foods concerned are ready-to-eat foods that promote the growth of the pathogen during their shelf lives. The main food groups considered were cold meats, soft cheeses with raw or pasteurized milk, cold smoked fish, shellfish, raw vegetables and mixed foods. The duration and temperature of preservation during storage are therefore essential factors to be monitored.
Shiga toxin-producing *E. coli*	Foods associated with STEC infections include minced beef, raw milk, soft cheeses made from raw milk and products eaten raw. Minced beef can be implicated when it is eaten raw (tartare) or undercooked. The only potential actions at the preparation stage of dairy products are heating the raw milk or avoiding consumption of raw milk products, especially for young children who demonstrate greater host sensitivity.
Non-typhoid *Salmonella* enterica	The main foods considered were minced beef, chicken and pork meats, cheeses made with raw milk, raw eggs and egg products and raw products. The favorable conditions for handling such food (consumed raw or undercooked) and the action strategies (cooking, washing and disinfection, disposal) are the same as with STEC for meat-based products, cheeses, fruits and vegetables. Improper storage is also considered to be an important factor for salmonellosis associated with meats. Infections connected with eggs are linked to undercooking, and those linked to raw egg-based products are sometimes associated with cross-contamination and improper storage temperature.
Norovirus	Foodborne transmission of norovirus is connected with bivalve mollusks, raw products (frozen or non-frozen) and meals involving extensive handling, such as sandwiches. Hoffmann *et al.* (2007) and Batz *et al.* (2012) identified fruit and vegetables and seafood as the main foods responsible for the transmission of norovirus in the United States.

	Cooking bivalve mollusks is effective in destroying viral particles. As for complex foods that are handled, contamination usually occurs during food handling through cross-contamination.

Table A.4. *Short presentation of the hazards listed in Table A.3 (Augustin* et al. *2020)*

References

Augustin, J.-C., Kooh, P., Bayeux, T., Guillier, L., Meyer, T., Jourdan-Da Silva, N., Villena, I., Sanaa, M., Cerf, O., Blanchemanche, S., Bonnaud, L., Gautier, M., Gauchard, F., Nabec, L., Soler, L.-G., ANSES Working Group (2020). Contribution of foods and poor food-handling practices to the burden of foodborne infectious diseases in France. *Foods*, 9(11), 1644–1661.

Batz, M.B., Hoffmann, S., Morris Jr., J.G. (2012). Ranking the disease burden of 14 pathogens in food sources in the United States using attribution data from outbreak investigations and expert elicitation. *Journal of Food Protection*, 75(7), 1278–1291.

Beumer, R.R., Te Giffel, M.C., De Boer, E., Rombouts, F.M. (1996). Growth of *Listeria monocytogenes* on sliced cooked meat products. *Food Microbiology*, 13(4), 333–340.

Braithwaite, P.J., Perigo, J.A., Barker, A.M., Gould, G.W., Skinner, F.A. (1971). The influence of of pH, water activity and recovery temperature on the resistance and outgrowth of *Bacillus* spores. In *Spore Research*, Barker A.M., Gould G.W., Wolf, J. (eds). Academic Press, London.

Carlier, V., Augustin, J.C., Rozier, J. (1996). Destruction of *Listeria monocytogenes* during a ham cooking process. *Journal of Food Protection*, 59(6), 592–595.

Carlin, F., Broussolle, V., Perelle, S., Litman, S., Fach, P. (2004). Prevalence of *Clostridium botulinum* in food raw materials used in REPFED manufactured in France. *International Journal of Food Microbiology*, 91, 141–145.

Del Torre, M., Stecchini, M.L., Braconnier, A., Peck, M.W. (2004). Prevalence of *Clostridium* species and behaviour of *Clostridium botulinum* in gnocchi, a REPFED of Italian origin. *International Journal of Food Microbiology*, 96, 115–131.

EFSA (2014). Trends and sources of zoonoses and zoonotic agents in foodstuffs, animals and feedingstuffs in 2012. *EFSA Journal*, 12(2), 3547–3858.

Gombas, D.E., Chen, Y., Clavero, R.S., Scott, V.N. (2003). Survey of *Listeria monocytogenes* in ready-to-eat foods. *Journal of Food Protection*, 66(4), 559–569.

Gould, G.W. (1999). Sous vide foods: Conclusions of an ECFF botulinum working party. *Food Control*, 10(1), 47–51.

Hauschild, A.H.W. and Simonsen, B. (1985). Safety of shelf-stable canned cured meats. *Journal of Food Protection*, 48, 997–1009.

Havelaar, A.H., Haagsma, J.A., Mangen, M.-J.J., Kemmeren, J.M., Verhoef, L.P.B., Vijgen, S.M.C., Wilson, M., Friesema, I.H.M., Kortbeek, L.M., Van Duynhoven, Y.T.H.P., Van Pelt, W. (2012). Disease burden of foodborne pathogens in the Netherlands, 2009. *International Journal of Food Microbiology*, 156(3), 231–238.

Hoffmann, S., Fischbeck, P., Krupnick, A., McWilliams, M. (2007). Using expert elicitation to link foodborne illnesses in the United States to foods. *Journal of Food Protection*, 70(5), 1220–1229.

Hwang, C.A., Huang, L.H., Sheen, S., Juneja, V. (2012). Effects of lactic acid on the growth characteristics of *Listeria monocytogenes* on cooked ham surfaces. *Journal of Food Protection*, 75(8), 1404–1410.

Keskinen, L.A., Todd, E.C.D., Ryser, E.T. (2008). Transfer of surface-dried *Listeria monocytogenes* from stainless steel knife blades to roast turkey breast. *Journal of Food Protection*, 71(1), 176–181.

Little, C.L., Sagoo, S.K., Gillespie, I.A., Grant, K., McLauchlin, J. (2009). Prevalence and level of *Listeria monocytogenes* and other *Listeria* species in selected retail ready-to-eat foods in the United Kingdom. *Journal of Food Protection*, 72(9), 1869–1877.

Lund, B.M. and Peck, M.W. (2000). *Clostridium botulinum*. In *The Microbiological Safety and Quality of Food*, Lund, B.M., Baird-Parker, T.C., Gould, G.W. (eds). Aspen Publishers, Gaithersburg.

Martins, E.A. and Germano, P.M.L. (2011). *Listeria monocytogenes* in ready-to-eat, sliced, cooked ham and salami products, marketed in the city of Sao Paulo, Brazil. Occurrence, quantification, and serotyping. *Food Control*, 22(2), 297–302.

Mataragas, M., Drosinos, E.H., Siana, P., Skandamis, P., Metaxopoulos, I. (2006). Determination of the growth limits and kinetic behavior of *Listeria monocytogenes* in a sliced cooked cured meat product: Validation of the predictive growth model under constant and dynamic temperature storage conditions. *Journal of Food Protection*, 69(6), 1312–1321.

Membré, J.-M., Wemmenhove, E., McClure, P.J. (2009). Exposure assessment model to combine thermal inactivation (log reduction) and thermal injury (heat-treated spore lag time) effects on non-proteolytic *Clostridium botulinum*. In *Safety, Reliability and Risk Analysis: Theory, Methods and Applications*, Martorell, S., Soares, C.G., Barnett, J. (eds). CRC Press, London.

Murray, C.J. and Acharya, A.K. (1997). Understanding DALYs (disability-adjusted life years). *Journal of Health Economics*, 16(6), 703–730.

Notermans, S., Dufrenne, J., Lund, B.M. (1990). Botulism risk of refrigerated, processed foods of extended durability. *Journal of Food Protection*, 53, 1020–1024.

Peck, M.W. (1997). *Clostridium botulinum* and the safety of refrigerated processed foods of extended durability. *Trends in Food Science and Technology*, 8, 186–192.

Peck, M.W., Evans, R.I., Fairbairn, D.A., Hartley, M.G., Russel, N.J. (1995). Effect of sporulation temperature on some properties of spores of non-proteolytic *Clostridium botulinum*. *International Journal of Food Microbiology*, 28, 289–297.

Pivnick, H. and Petrasovits, A. (1973). A rationale for the safety of canned shelf-stable cured meats: Protection = destruction + inhibition. *19th Meeting European Meat Researchers*. Paris, France.

Sheen, S. and Hwang, C.A. (2008). Modeling transfer of *Listeria monocytogenes* from slicer to deli meat during mechanical slicing. *Foodborne Pathogens and Disease*, 5(2), 135–146.

U.S. Food and Drug Administration (2001). Fish and fisheries products hazards and controls guidance [Online]. Available at: http://www.cfsan.fda.gov/~comm/haccp4.html.

Uyttendaele, M., De Troy, P., Debevere, J. (1999). Incidence of *Listeria monocytogenes* in different types of meat products on the Belgian retail market. *International Journal of Food Microbiology*, 53(1), 75–80.

WHO (2017). The burden of foodborne diseases in the WHO European Region [Online]. Available at: https://www.euro.who.int/data/assets/pdffile/0005/402989/50607-WHO-Food-Safety-publicationV4_Web.pdf.

Zhang, L., Moosekian, S.R., Todd, E.C.D., Ryser, E.T. (2012). Growth of *Listeria monocytogenes* in different retail delicatessen meats during simulated home storage. *Journal of Food Protection*, 75(5), 896–905.

Zwietering, M.H., Jacxsens, L., Membré, J.-M., Nauta, M.J., Peterz, M. (2016). Relevance of microbial finished product testing in food safety management. *Food Control*, 60, 31–43.

Glossary

Codex Alimentarius: A collection of standards, guidelines and codes of practice adopted by the Codex Alimentarius Commission.

Codex Alimentarius Commission (CAC): The central part of the Joint FAO/WHO Food Standards Programme. It was established by the FAO and WHO to protect consumer health and promote fair practices in food trade.

Dose–response assessment: The determination of the relationship between the magnitude of exposure (dose) to a chemical, biological or physical agent and the severity and/or frequency of associated adverse health effects (response).

Exposure assessment: The qualitative and/or quantitative evaluation of the likely intake of biological, chemical and physical agents via food, as well as exposures from other sources if relevant.

Food and Agriculture Organization (FAO): A specialized agency of the United Nations, the goal of which is to achieve food security for all and make sure that people have regular access to enough high-quality food to lead active, healthy lives.

Food safety: The assurance that a food will not cause harm to the consumer when prepared and/or consumed in accordance with its intended use.

Hazard: A biological, chemical or physical agent in or a condition of food with the potential to cause an adverse health effect.

Hazard characterization: The qualitative and/or quantitative evaluation of the nature of the adverse health effects associated with the biological, chemical and physical agents that may be present in food.

Hazard identification: The identification of biological, chemical and physical agents capable of causing adverse health effects and which may be present in a particular food or group of foods.

Predictive microbiology: Scientific approach based on mathematical models aiming to predict the behavior of microbial populations in foods or food model systems over time as a function of extrinsic factors, such as storage temperature, and intrinsic factors, such as pH and water activity. Predictive microbiology and exposure assessment are closely connected: predictive microbiology enables the quantification of the level of microorganisms in a consumer's portion, information required for exposure assessment.

Qualitative risk assessment: A descriptive approach which, when conditioned by prior expert knowledge and identification of attendant uncertainties, allows risk ranking or separation of risks into descriptive categories.

Quantitative risk assessment: A risk assessment that provides numerical expressions of risk and indications of the attendant uncertainties.

Risk: A function of the probability of an adverse health effect and the severity of that effect as a consequence of a hazard or hazards in food.

Risk analysis: A process consisting of three components – risk assessment, risk management and risk communication.

Risk assessment: A scientifically based process consisting of the following steps: (i) hazard identification, (ii) hazard characterization, (iii) exposure assessment and (iv) risk characterization.

Risk characterization: The process of producing a qualitative and/or quantitative estimation, including attendant uncertainties, of the probability of occurrence and severity of known or potential adverse health effects in a given population based on hazard identification, hazard characterization and exposure assessment.

Risk communication: The interactive exchange of information and opinions concerning risks, risk-related factors and risk perceptions throughout the risk analysis process among risk assessors, risk managers, consumers, industry actors, the academic community and other interested parties, including the explanation of risk assessment findings and of the basis of risk management decisions.

Risk estimate: The qualitative and/or quantitative estimation of risk resulting from risk characterization.

Risk management: The process, distinct from risk assessment, of weighing policy alternatives in consultation with all interested parties, considering risk assessment and other factors relevant for the health protection of consumers and the promotion of fair-trade practices, and, if needed, selecting appropriate prevention and control options.

Semi-quantitative risk assessment: Semi-quantitative risk assessment involves assigning numbers to qualitative estimates of exposure and dose–response relationships, in the form of probability ranges, weights or scores, and combining them through addition, multiplication or other mathematical operations to arrive at a risk estimate, with the objective of achieving a greater level of objectivity compared to a qualitative risk assessment approach.

Sensitivity analysis: a method used to examine the behavior of a model by measuring the variation in its outputs resulting from changes to its inputs.

Uncertainty analysis: A method used to estimate the uncertainty associated with model inputs, assumptions and structure/form.

World Health Organization (WHO): A specialized agency of the United Nations responsible for international public health.

List of Authors

Mayra Aguirre GARCIA
SECALIM
INRAE
Oniris
Nantes
France

Laurent GUILLIER
ANSES
Department of Risk Assessment
Maison-Alfort
France

Nabila HADDAD
SECALIM
INRAE
Oniris
Nantes
France

Jeanne-Marie MEMBRÉ
SECALIM
INRAE
Oniris
Nantes
France

Maarten NAUTA
Statens Serum Institut
Copenhagen
Denmark

Moez SANAA
ANSES
Department of Risk Assessment
Maison-Alfort
France

Index

G, H, I

L, M

N, P

Q, R

S, T

U, V, W

Printed and bound by CPI Group (UK) Ltd, Croydon, CR0 4YY
13/09/2026
14960995-0002